Soma and the Indo-European Priesthood

Soma and the Indo-European Priesthood

William Scott Shelley

Algora Publishing
New York

Library of Congress Cataloging-in-Publication Data —

Names: Shelley, William Scott, author.
Title: Soma and the Indo-European priesthood: cereal cultivation and the origins of religion / William Scott Shelley.
Description: New York: Algora Publishing, 2018. | Includes bibliographical references.
Identifiers: LCCN 2018043269 (print) | LCCN 2018045556 (ebook) | ISBN 9781628943535 (pdf) | ISBN 9781628943511 (soft cover: alk. paper) | ISBN 9781628943528 (hard cover: alk. paper)
Subjects: LCSH: Indo-Europeans—Religion. | Grain—Religious aspects.
Classification: LCC BL660 (ebook) | LCC BL660 .S445 2018 (print) | DDC 299/.1—dc23
LC record available at https://lccn.loc.gov/2018043269

Printed in the United States

Table of Contents

Chapter 1: The Indo-Europeans and Indo-Aryans

The Indo-Europeans are the various Caucasian cultures that extend from northern Europe to Central Asia and the Indian Ocean. The origin of the Indo-Europeans remains unsettled and is a question beyond the parameters of this monograph. Roughly twelve thousand years ago, when early man had already discovered fire and was beginning to develop pottery and agriculture, a revolution had begun that would lead to the development of the great civilizations of Egypt, Mesopotamia, Persia, India, Thrace, Greece, and Rome.

This revolution was the cultivation of cereals, and it would come to be called the "Agricultural Revolution" of the Old World. The development of agriculture curtailed the nomadic lifestyle of the hunter and gatherer and replaced this with a sedentary one. The inhabitants of the Old World abandoned the nomadic life in favor of a settled existence, after which small villages grew into towns, and then city-states, bringing with this greater social stratification.

During this early period in history, before there was the development of a priestly class, religion was the domain of the medicine man, the shaman. In fact, all religion originated with the medicine man, and from these early shamans there developed the priesthood and religion of the Indo-Europeans.[1] As the medicine man employed plants, so too did the priesthoods of the

[1] Weston La Barre, "The Deathless Gods," *The Ghost Dance: Origins of Religion*, New York, Delta Publishing Co. Inc., 1970, pp. 433-476.

Indo-European people. The Brahman priests of India utilized a plant they called Soma, a plant identical to what the Aryans of Persia (modern Iran, a name derived from Aryan) named Haoma, the Chaldeans of the Arabs called Lotus, the priests of the Thracians and Greeks named Moly, the Druid priests of Britain named Mistletoe, and the priests of Egypt named Hyssop.[1] The identity of the Indo-Aryan Soma/Haoma plant has generated a number of theories, including the one contained within these pages.

The Indo-Aryans are a branch of the Indo-Europeans, which includes those who migrated out of greater Iran into the region of the Indus River (modern Pakistan), the people of the Vedas. Within these sacred texts the priests speak of a supernatural plant called Soma, which the Vedic texts make central to the religion, and the Vedic religion was central to the culture. This sacred drink was also called "mead" by these priests, reminiscent of the mead of Odin (Wotan) of the Germanic people, who are also related to this branch of the Indo-Europeans.

The Brahman priests of South Asia were performing the Soma rites that were brought from the land of the Aryan Magi, whose rites were reformed by Zarathustra, or Zoroaster (*ca.* seventh century B.C.). These same Magi were present at the birth of Jesus, bearing herbal gifts. The Greek historian Strabo says that the Greek historian Megesthenes reports that there are two types of philosophers in India, the Brahmans and the Gymnosophists ("naked priests").[2] According to the Greek philosopher Porphyry, the former was derived from one stock, and the latter was "collected from every nation of Indians."[3] The Christian theologian Clement of Alexandria writes the following of the Gymnosophists in India:

> There are two classes of these, called Sarmans and Brahmans. Among the Sarmans, the so-called forest dwellers do not occupy cities or have roofs over their heads. They wear tree bark, take their food from berries and drink water from their hands. They do not recognize marriage or the procreation of children, like our present day so-called Encratites (an ascetic Christian sect). Among the Indians are some who follow the precepts of Buddha, whom for his

[1] Porphyry, *Abstinence from Animal Food* IV.6.
[2] Strabo, *Geographica* XV.1.59.
[3] Porphyry, *Abstinence from Animal Food* IV.17.

exceptional sanctity they have honored him as a god.[1]

Another great reformer of the Soma cult, although he failed, was Gautama Siddhartha (*ca.* fifth century B.C.), also known as Gautama Buddha. Gautama was born to royalty in Magadha, a region of India that the Aryans had settled centuries earlier, emerging as the dominant military power, and consequently the established monarchy of the region. Destined to become either a great king or the Buddha, Gautama chose the latter, a more difficult and dangerous path. Not only did Gautama employ cannabis seeds on his journey, his enlightenment was directly associated with the Soma plant that the priests who inhabited the region were consuming on a regular basis. If Gautama had chosen to become king, these same Brahman priests would have been responsible for consecrating the prince with a ritual that involved the consumption of Soma by both the priests and the prospective king.

In the Pali literature, the bohdi "fig" tree under which Gautama Siddhartha gained enlightenment and became the Buddha ("Awake") is named nygrodha and assvattha ("horse station"). The nygrodha is not mentioned once in the *Ṛgveda*, and the assvattha was sacred in the Indus valley prior to the appearance of the Vedic Aryans in the region. In the *Atharvaveda*, the gods pronounced the *kushtha*, "the visible manifestation of amrita (ambrosia)," at the assvatha tree.[2] Furthermore, in the *Chandogyo Upanishad*, the assvattha tree is said to drip Soma.[3]

According to the Greek historian Herodotus, the Magi were a tribe of Medes.[4] Herodotus also says that the Medes "were in old time called by all men Arians, but when the Colchian woman Medea came from Athens among the Arians they changed their name, like the Persians. This is the Medes' own account of themselves."[5] The Greek geographer Pausanias concurs and says Medea fled from Athens, and "Coming to the land then called Aria, she caused its inhabitants to be named after her Medes."[6] Herodotus describes the Colchians as a dark-skinned, wooly-haired people, and they occupied

[1] Clement of Alexandria, *Stromata* I.71.

[2] *Atharvaveda* V.4.1-6; in K. L. Joshi (ed.), *Atharva-Veda Sa§hitā*, Volume I, Delhi, Parimal Publications, 2000, pp. 374-383.

[3] *Chandogyo Upanishad* VIII.5.3; in K. T. Pandurangi (trans.), *Chandogyopanishad*, Chirtanur, Sriman Madhua Siddhantonnahini Sabha, 1987, p. 341.

[4] Herodotus, *Histories* I.101.

[5] Ibid. VII.62.

[6] Pausanias, *Description of Greece* II.3.8.

the Phasis River in Asia Minor that flowed into the Euxine Sea (Black Sea), dividing Asia from Europe, and some of these later settled the Italian city of Polae.[1] The Colchians were represented in the army of Sesostris (Ramses II, 14th century B.C.), the only Egyptian king to also rule Ethiopia; after subduing Asia and Europe, he left behind the Colchians who settled near the Phasis River.[2]

According to the Greek historian Athenaeus, the Magi were said to have incestual relations with their mothers, sisters, and daughters.[3] Clement writes, "Xathus in his book entitled the *Works of the Magi* says, 'The Magi think it right to have sexual union with their mothers, daughters and sisters. The women are held in common by mutual agreement, not forcibly or secretively, when one wants to marry another's wife.'"[4] The Latin historian Aelian writes:

> The wisdom of the Persian Magi was (besides other things proper to them) conversant in prediction; they foretold the cruelty of Ochus towards his subjects, and his bloody disposition, which they collected from some secret signs. For when Ochus, upon the death of his father Artaxerxes, came to the crown, the Magi charged one of the eunuchs that were next to him to observe upon what things, when the table was set before him, he first laid hands; who watching intentively, Ochus reached forth both hands, and with his right hand took hold of a knife that lay by, and with the other took a great loaf, which he laid upon the meat, and did cut and eat greedily. The Magi, hearing this, foretold that there would be plenty during his reign, and much bloodshed. In which they erred not.[5]

In the fourth century B.C., the Greek historian Xenophon records in his *Cyropaedia*, the future King Cyrus' father Cambyses, king of the Persians, encourages his son to learn the ways of the Magi, he says to Cyrus: "So we see that mere human wisdom does not know how to choose what is best...But the gods, my son, the eternal gods, know all things, both what has been and what is and what shall come to pass as a result of each present or past event;

[1] Herodotus, *Histories* II.104; Strabo, *Geographica* I.2.39; XI.2.18; Procopius, *Buildings* VI.1.7.
[2] Herodotus, *Histories* II.103-110.
[3] Athenaeus, *Deipnosophistae* V.220.
[4] Clement of Alexandria, *Stromata* III.11.
[5] Aelian, *Various History* II.17.

and if men consult them, they reveal to those to whom they are propitious what they ought to do and what they ought not to do."[1] Xenophon writes:

> Now, when Cyrus had gone home and prayed to ancestral Hestia, ancestral Zeus, and the rest of the gods, he set out upon his expedition; and his father also joined in escorting him on his way. And when they were out of the house, it is said to have thundered and lightened with happy auspices for him; and when this manifestation had been made, they proceeded, without taking any further auspices, in the conviction that no one would make void the signs of the supreme god. Then, as they went on, his father began to speak to Cyrus in this wise: "My son, it is evident both from the sacrifices and from the signs from the skies that the gods are sending you forth with their grace and favour; and you yourself must recognize it, for I had you taught this art on purpose that you might not have to learn the counsels of the gods through others as interpreters, but that you yourself, both seeing what is to be seen and hearing what is to be heard, might understand; for I would not have you at the mercy of the soothsayers, in the case they should wish to deceive you by saying other things than those revealed by the gods; and furthermore, if you should be without a soothsayer, I would not have you in doubt as to what to make of the divine revelations, but by your soothsayer's art I would have you understand the counsels of the gods and obey them."[2]

According to Xenophon, Cyrus "never failed to sing hymns to the gods at daybreak and to sacrifice daily to whatsoever deities the Magi directed…[T] he rest of the Persians also imitated him from the first; for they believed that they would be more sure of good fortune if they revered the gods just as he did who was the sovereign and most fortunate of all…"[3] Xenophon says that Cyrus (560–529 B.C.) was visited in a dream and was told of his impending death. "As he slept in the palace, he saw a vision: a figure of more than human majesty appeared to him in a dream and said, 'Be packing up Cyrus; for thou shalt soon depart to the gods.' When the vision was past, he awoke

[1] Xenophon, *Cyropaedia* I.6.46.
[2] Ibid. I.6.1-2.
[3] Ibid. VIII.1.23-24.

and seemed almost to know that the end of his life was at hand."[1]

Xenophon records the deterioration of the character, religion, culture, and civilization of the Persians following the death of Cyrus:

> That Cyrus's empire was the greatest and most glorious of all the kingdoms of Asia—of that it may be its own witness. For it was bounded on the east by the Indian Ocean, on the north by the Black Sea, on the west by Cyprus and Egypt, and on the south by Ethiopia...as soon as Cyrus was dead, his children at once fell into dissension, states and nations began to revolt, and everything began to deteriorate. And that what I say is the truth, I will prove, beginning with the Persians' attitude toward religion...Witnessing such a state of morality, all of the inhabitants of Asia have turned to wickedness and wrongdoing. For, whatever the character of the rulers is, such also that of the people under them for the most part becomes...In money matters, too, they are more dishonest...Now, to be sure, the custom of eating but once a day still prevails, but they begin to eat at the hour when those who have breakfast earliest begin their morning meal, and they keep on eating and drinking until the hour when those who stay up latest go to bed...at their banquets they drink so much that...they are themselves carried out when they are no longer able to stand straight enough to walk out. Again, this also was a native custom of theirs, neither to eat nor drink while on a march, nor yet to be seen doing any of the necessary consequences of eating or drinking. Even yet that same abstinence prevails, but they make their journeys so short that no one would be surprised at their ability to resist those calls of nature... Artaxerxes and his court became the victims of wine...instruction and practice of horsemanship have died out...And while anciently the boys there used to hear cases of law justly decided and so learn justice, as they believed—that also has been entirely reversed; for now they see all too clearly that whichever party gives the larger bribe wins the case. The boys of that time used also to learn the properties of the products of the earth, so as to avail themselves of the useful ones and keep away from those that were harmful. But

[1] Ibid. VIII.7.2.

> now it looks as if they learned them only in order to do as much harm as possible; at any rate, there is no place where more people die or lost their lives from poisons than there. Furthermore, they are much more effeminate now than they were in Cyrus's day...they are allowing the rigour of the Persians to die out, while they keep up the effeminacy of the Medes...I maintain that I have proved that the Persians of the present day and those living in their dependencies are less reverent towards the gods, less dutiful to their relatives, less upright in their dealings with all men, and less brave in war than they were of old.[1]

Xenophon records the events that took place when the young Cyrus visited his grandfather Astyages, king of Media:

> Astyages dined with his daughter and Cyrus...Now the cupbearers of those kings...pour in the wine with neatness and then present the goblet...in such a way as to place it most conveniently in the grasp of the one who is to drink...[Cyrus later said], "I saw that you were unsteady both in mind and in body. For in the first place you yourselves kept doing what you never allow us boys to do; for instance, you kept shouting, all at the same time, and none of you heard anything that the others were saying; and you fell to singing, and in a most ridiculous manner at that, and though you did not hear the singer, you swore that he sang most excellently; and though each one of you kept telling stories of his own strength, yet if you stood up to dance, to say nothing of dancing in time, why, you could not even stand up straight. And all of you quite forgot—you, that you were king; and the rest, that you were their sovereign. It was then that I also for my part discovered, and for the first time, that what you were practising was your boasted 'equal freedom of speech'; at any rate, never were any of you silent." "But my boy," Astyages said, "does not your father get drunk, when he drinks?" "No, by Zeus," said he. "Well, how does he manage it?" "He just quenches his thirst and thus suffers no further harm..."[2]

According to the Greek orator Isocrates, the Persians and Scythians

[1] Ibid. VIII.8.1-27.
[2] Ibid. I.3.4-11.

possessed a strong instinct for domination: "Let us single out, then, the races which have the strongest instinct for domination and the greatest power of aggression—the Scythians and the Thracians and the Persians."[1] According to Herodotus, in Europe beyond the Ister (Danube) dwelt the Sigynnae, who wore Median clothes and called themselves colonists from Persian Media.[2] Strabo writes: "The Siginni imitate the Persians in all their customs."[3] Referring to the Scythians, the Latin historian Pliny writes: "the Persians have given the general name Sacae for the tribe nearest to Persia, and the Scythians themselves," adding that "the name of Scythians has spread in every direction, as far as the Sauromatae and the Germans, but this old designation has not continued for any except the most outlying sections of these races…"[4]

In the fifth century B.C., the Greek physician Hippocrates locates the Scythian people named Sauromatae around the Sea of Azov: "And in Europe is a Scythian race, dwelling round Lake Maeotis (Sea of Azov), which differs from the other races. Their name is Sauromatae."[5] Herodotus writes, "Above the Royal Scythians to the north dwell the Blackcloaks (Melanchlaeni)…and beyond the Blackcloaks the land is all marshes and uninhabited by men, so far as we know," adding, "The Blackcloaks all wear black raiment, whence they take their name; their usages are Scythian."[6] The Byzantine historian Procopius writes in the sixth century A.D.:

There were many Gothic nations in earlier times, just as also at the present, but the greatest and most important of all are the Goths, Vandals, Visigoths, and Gepaedes. In ancient times, however, they were named Sauromatae and Melanchlaeni; and there were some too who called these nations Getic. All these, while they have been distinguished from one another by their names, as has been said, do not differ in anything else at all. For they all have white bodies and fair hair, and are tall and are handsome to look upon, and they use the same laws and practise a common religion. For they are all of the Arian faith, and have one language called Gothic; and, as it seems to me, they all

[1] Isocrates, *Panathenaicus* 67.
[2] Herodotus, *Histories* IV.6-7.
[3] Strabo, *Geographica* XI.11.8.
[4] Pliny, *Natural History* VI.50; VI.81.
[5] Hippocrates, *Airs Waters Places* 17.
[6] Herodotus, *Histories* IV.20; IV.107.

came originally from one tribe, and were distinguished later by the names of those who led each group. This people used to dwell above the Ister River (Danube) from of old.[1]

Herodotus says that the Scythians own account of themselves was that they originally appeared in the northern territory of Asia one thousand years before Darius, or *ca.* 1500 B.C., around the very time that the Indo-Aryans were migrating into the Indus valley: "Such then is the Scythians' account of their origin; they reckon that neither more nor less than a thousand years in all passed between their first appearing and the crossing over of Darius into their country."[2] According to the Roman Emperor Julius Caesar, the German tribes were completely without Druid priests, or any priesthood.[3]

In the ninth century B.C., the Greek poet Homer writes in the *Iliad*: "but himself turned away his bright eyes, and looked afar, upon the land of the Thracian horsemen, and of the Mysians that fight in close combat, and of the Hippemogi that drink the milk of mares, and of the Abii, the most righteous of men."[4] Strabo identifies the latter two cultures as Scythian and Sauromatae: "the 'Hipppemolgi' and 'Abii' are indeed the wagon-dwelling Scythians and Sarmatians. For at the present time these tribes, as well as the (German) Bastarnian tribes, are mingled with the Thracians (more indeed with those outside the Ister, but also with those inside). And mingled with them also are the Celtic tribes—the Boii, the Scordisci and the Taurisci."[5]

Included among the Indo-Europeans are the Thraco-Phrygians of Southeast Europe and Asia Minor. After traveling to Thrace in the fifth century B.C., Herodotus writes: "The Thracians are the biggest nation in the world, next to the Indians; were they under one ruler, or united, they would in my judgment be invincible and the strongest nation on earth; but since there is no way or contrivance to bring this about, they are for this reason weak. They have many names, each tribe according to its region."[6] The Thracians were reputed to be the "fiercest of races," making Thrace "one of the most powerful nations of Europe."[7]

[1] Procopius, *History of the Wars* III.2.1-6.
[2] Herodotus, *Histories* V.9.
[3] Julius Caesar, *Gallic War* VI.21.
[4] Homer, *Iliad* XIII.4-8.
[5] Strabo, *Geographica* VII.3.2.
[6] Herodotus, *Histories* V.3.
[7] Pliny, *Natural History* IV.40; Velleius Paterculus, *Roman History* II.98.2.

According to Isocrates, the Thracians possessed an instinct for domination and power of aggression.[1] Homer says that the horse-tending Thracians were experts in hand-to-hand fighting.[2] In the surviving fragments of the Greek poet Archilochus, the soldier-poet calls the Thracians the gods of battle, and confesses in *The Poet's Shield* that he once fled the battlefield from the Thracians, leaving his shield behind as a war trophy.[3] Among all of his troops, Alexander the Great called the Thracian soldiers "the stoutest in Europe, and the most warlike."[4] Speaking of the Thracians, Herodotus writes:

> Now whether this separation, like other customs has come to Greece from Egypt, I cannot exactly judge. I know that in Thrace and Scythia and Persia and Lydia and nearly all foreign countries those who learn trades are held in less esteem than the rest of the people, and those who have the least to do with artisans' work, especially men who are free to practise the art of war, are highly honoured. This much is certain, that this opinion, which is held by all Greeks and chiefly by the Lacedaemonians, is of foreign origin. It is in Corinth that artisans are held in least contempt.[5]

According to Homer, the Thracian men wore their hair long.[6] Elsewhere the Thracians were described as red-haired, and "light-haired and grey-eyed."[7] The Thracians also practiced the custom of tattooing the body, injecting the design with pins.[8] Herodotus writes:

> Among the rest of the Thracians, it is the custom to sell their children to be carried out of the country. They take no care of their maidens, allowing them to have intercourse with what man they will; but their wives they strictly guard, and buy them for a great price from the parents. To be tattooed is a sign of noble birth; to bear no such marks is for the baser sort. The idler is most honoured, the tiller of the soil most contempted; he is held in highest honour who

[1] Isocrates, *Panathenaicus* 67.
[2] Homer, *Iliad* VIII.4-8.
[3] Archilochus, *The Poet's Shield* 1-8.
[4] Arrian, *Anabasis of Alexander* II.7.5.
[5] Herodotus, *Histories* II.167.
[6] Homer, *Iliad* IV.533.
[7] Clement of Alexandria, *Stromata* VII.22; Firmicus Maternus, *Mathesis* I.1.
[8] Dio Chrysostom, *Discourses* XIV.18; Athenaeus, *Deipnosophistae* XII.524; Phanocles, *Fragment* 1-4.

> lives by war and foray.[1]

Over the course of their life it was the custom of the Thracian men to take a dozen or more wives, a practice also followed by the Indians and the British.[2] Strabo quotes the Thracian poet Menander, a Getan who confessed the practice of polygamy by his people:

> And see the statement of Menander about them, which, as one may reasonably suppose, was not invented by him but taken from history: "All the Thracians, and most of all we Getae (for I too boast that I am of this stock) are not very continent"; and a little below he sets down the proofs of their incontinence in their relations with women: "For every man of us marries ten or eleven women, and some, twelve or more; but if anyone meets death before he has married more than four or five, he is lamented among the people there as a wretch without bride and nuptial song." Indeed, these facts are confirmed by the other writers as well.[3]

The Thracian Getae who lived north of the Danube called themselves immortals, and according to the Greek historian Arrian (2nd century A.D.), they were "the bravest and most law-abiding of all Thracians."[4] His contemporary, the historian Appian describes the Getae as a hardy and warlike nation.[5] The Getae crossed the Danube continuously, intermingling with the Thracians and the Thracians called Mysians.[6] The Daci and the Getae spoke the same language, and at one time these tribes attained great power.[7] These tribes not only counted the Romans as their enemies, but they also fought with the Germanic tribes, who were also the enemies of the Romans.[8]

According to the Greek philosopher Diogenes, the mystical figure named Zalmolxis was a Thracian who the Getae worshipped as the god Cronos.[9] According to Strabo, the Pythagorean doctrine of abstaining from animal flesh was taught by Zalmolxis, to which he adds:

---

[1] Herodotus, *Histories* V.6.
[2] Megesthenes, *Fragment* 27.
[3] Strabo, *Geographica* VII.3.4.
[4] Arrian, *Anabasis of Alexander* I.3.2-5; Herodotus, *Histories* IV.93.
[5] Appian, *The Civil Wars* II.110.
[6] Strabo, *Geographica* VII.3.10-13.
[7] Ibid. VII.3.13.
[8] Ibid.; Dio Cassius, *Roman History* LI.22.
[9] Diogenes Laertius, *Lives of Eminent Philosophers* I.1; VIII.2.

> In fact, it is said that a certain man of the Getae, Zalmolxis by name, had been a slave to Pythagoras, who had learned some things about the heavenly bodies from him, as also certain other things from the Egyptians, for in his wanderings he had gone even as far as Egypt; and when he came on back to his homeland he was eagerly courted by the rulers and the people of his tribe, because he could make predictions from the celestial signs; and at last he persuaded the king to take him as a partner in the government, on the ground that he was competent to report the will of the gods; and although at the outset he was only made a priest of the god who was most honoured in their country, yet afterwards he was even addressed as god, and having taken possession of a certain cavernous place that was inaccessible to anyone else he spent his life there, only rarely meeting with any people outside except the king and his own attendants; and the king cooperated with him, because he saw that the people paid much more attention to himself than before, and the belief that the decrees which he promulgated were in accordance with the counsel of the gods. This custom persisted even down to our own time, because some man of that character was always to be found, who, though in fact only a counselor to the king, was called god among the Getae.[1]

Herodotus held the belief that Zalmolxis had lived many years before Pythagoras: "Now the Thracians were a meanly-living and simple witted folk, but this Salmoxis knew Ionian usages and a fuller way of life than the Thracians; for he had consorted with Greeks," adding, "he made himself a hall, where he entertained and feasted the chief among his countrymen, and taught them that neither he nor his guests nor any of their descendants should ever die, but that they should go to a place where they would live for ever and have all good things."[2] According to Herodotus, every five years the Getae would sacrifice a member of their tribe, believing that the victim could thereby deliver their messages to Zalmolxis:

> [The Getae] believe that they do not die, but that he who perishes goes to the god Salmoxis, or Gebeleïzis as some of them call him.

[1] Strabo, *Geographica* VII.3.5.
[2] Herodotus, *Histories* IV.95.

> Once in every five years they choose by lot one of their people and send him as a messenger to Salmoxis, charged to tell of their needs, and this is their manner of sending: Three lances are held by men thereto appointed; others seize the messenger to Salmoxis by his hands and feet, and swing and hurl him aloft on to the spear-points.[1]

According to Strabo, the Getae continued to be obedient to the priests who came after Zalmolxis:

> Boerebistas a Getan, on setting himself in authority over the tribe, restored the people who had been reduced to an evil plight by numerous wars...To help him secure the complete obedience of his tribe he had as his coadjutor Decaeneus, a wizard, a man who not only had wandered through Egypt, but also had thoroughly learned certain prognostics through which he would pretend to tell the divine will; and within a short time he was set up as god (as I said when related the story of Zalmolxis). The following is an indication of their complete obedience: they were persuaded to cut down their vines and to live without wine. However, certain men rose up against Boerebistas and he was deposed before the Romans sent an expedition against him...[2]

According to Strabo, this phenomenon existed in many Indo-European cultures:

> For these things, whatever there may be in them, have at least been believed and sanctioned among men; and for this reason the prophets too were held in so much honour that they were deemed worthy to be kings...Such, also, were Amphiarāus, Trophonius, Orpheus, Musaeus, and the god among the Getae, who in ancient times was Zalmolxis, a Pythagoreian, and in my time was Decaeneus, the diviner of Boerebistas, and among the Bosporeni, Achaecarus; and, among the Indians, the Gymnosophists; and among the Persians, the Magi...Among the Assyrians, the Chaldeans; and among the Romans, the Tyrrhenian nativity-casters. Moses was such a person as these, as also his successors, who, with no bad beginning, turned

[1] Ibid. IV.94.
[2] Strabo, *Geographica* VII.3.11.

out for the worse.[1]

Quoting the historian Poseidonius as his authority, Strabo writes of the Thracians:

> [I]n accordance with their religion they abstain from eating any living thing, and therefore from their flocks as well; and that they use as food honey and milk and cheese, living a peaceable life, and for this reason are called both "god-fearing" and "capnobatae" [smoke-treaders]; and there are some of the Thracians who live apart from womankind; and these are called Ctistae [creators], and because of the honour in which they are held, have been dedicated to the gods and live with freedom from every fear...And of course to regard as both "god-fearing and capnobatae" those who are without women is very much opposed to the common notions on that subject: for all agree in regarding the women as the chief founders of religion, and it is the women who provoke the men to the more attentive worship of the gods, to festivals, and to supplications, and is a rare thing for a man who lives by himself to be found addicted to these things. See again what the same poet [Menander] says when he introduces as speaker the man who is vexed by the money spent by the women in connection with the sacrifices: "The gods are the undoing of us, especially us married men, for we must always be celebrating some festival"; and again when he introduces the Woman-hater, who complains about these very things: "we used to sacrifice five times a day, and seven female attendants would beat cymbals all round us, while others would cry out to the gods."[2]

According to the Greek orator Demosthenes, unlike their German neighbors, "killing one another is not customary among the Thracians."[3] However, over time some of the Thracian tribes degenerated into a life of violence, sexual promiscuity, and even pederasty, including the tribe that Orpheus was said to have descended from, as the Latin historian Ammianus (4th century A.D.) writes: "The Odrysae are noted for the savage cruelty beyond all others, being so habituated to the shedding of human blood that

[1] Ibid. XVI.2.39.
[2] Ibid. VII.3.3-4.
[3] Demosthenes, *Against Aristocrates* 169.

when there were no enemies at hand, at their feasts, after a satiety of food and drink they plunged the sword into the bodies of their own countrymen, as if they were those of foreigners."[4] According to his contemporary, the poet Ausonius of Bordeaux, the Thracians in his time were "a race of uncurbed folk, with whom every crime is right—right bids them offer men in sacrifice..."[5] During his life, Procopius (6th century A.D.) witnessed the decline, and "practically all the women had become corrupt in character. For they sinned against their husbands with complete license...many of the adulterers actually attained honour from this conduct..."[6] During the lifetime of Procopius the Thracians suffered from internal and external violence, high crime rates, and moral degeneration, all of which contributed to the fall of the Eastern Roman Empire in the middle of the first millennium A.D. According to Procopius, these tribes even began to imitate the fashions of their enemies:

> In the first place, the mode of dressing the hair was changed to a rather novel style by the Factions; for they did not cut it at all as the other Romans did. For they did not touch the moustache or the beard at all, but they wished always to have the hair of these grow out very long, as the Persians do. But the hair of their heads they cut off in front back to the temples, leaving the part behind the head down to a very great length in a senseless fashion, just as the Massagetae do. Indeed for this reason they used to call this the "Hunnic" fashion.[7]

The Emperor Justinian ruled the Eastern Roman Empire during this time from Byzantium (Constantinople, modern Istanbul). There were many who considered him to be a demon, and along with his former prostitute wife, they both managed to destroy the nations of Thrace and the empire, all under the guise of Christianity. In his *Secret History* of Justinian's reign of crime and terror, Procopius writes:

> And while he [Justinian] seemed to have a firm belief as regards Christ, yet even this was for the ruin of his subjects. For he permitted the priests with comparative freedom to outage their neighbors, and if they plundered the property of the people whose land adjoined

[4] Ammianus Marcellinus, *The Chronicles of Events* XXVII.4.9.
[5] Ausonius of Bordeaux, *Technopaegnion* III.7-10.
[6] Procopius, *Secret History* XVII.24-26.
[7] Ibid. VII.8-10.

> theirs, he would congratulate them, thinking that thus he was showing reverence to the Deity. And in adjudicating such cases, he considered that he was acting in a pious manner if any man in the name of religion succeeded by his argument in seizing something that did not belong to him, and having won the case, went his way. For he thought that justice consisted in the priests' prevailing over their antagonists. And he himself, upon acquiring by means which were entirely improper the estates of persons either living or deceased and immediately dedicating them to one of the Churches, would feel pride in this pretence of piety, his object, however, being that title in these estates should not revert to the injured owners. Nay, more, he carried out an indefinite number of murders to accomplish these ends. For in his eagerness to gather all men into one belief as to Christ, he kept destroying the rest of mankind in a senseless fashion, and that too while acting with pretence of piety. For it did not seem to him murder if the victims chanced to be not of his own creed.[1]

The Thracian Orpheus was a shamanistic-bard who employed music by means of singing to a lyre, mystic rites, soothsaying, and also he is said to have discovered cures for various diseases.[2] Concerning Orpheus, Pausanias writes, "There are many untruths believed by the Greeks, one of which is that Orpheus was a son of the Muse Calliopê, and not of the daughter of Pierus, that the beasts followed him fascinated by his songs, and that he went down alive to Hades to ask for his wife from the gods below."[3] Pausanias records a story that circulated in Greece that involved a prophesy about the bones of Orpheus:

> The Macedonians who dwell in the district below Mount Pieria and the city of Dium say that it was here that Orpheus met his end at the hands of the women. Going from Dium along the road to the mountain, advancing twenty stades, you come to a pillar on the right surmounted by a stone urn which according to the natives contains the bones of Orpheus...In Larissia I heard another story, how that on

[1] Ibid. XIII. 4-7.

[2] Pausanias, *Description of Greece* VI.20.18, X.7.2; Plato, *Protagoras* 316D; Ovid, *Metamorphoses* XI.1-20.

[3] Pausanias, *Description of Greece* IX.30.4.

> Olympus is a city Libethra, where the mountain faces, Macedonia, not far from which city is the tomb of Orpheus. The Libethrians, it is said, received out of Thrace an oracle from Dionysus, stating that when the sun should see the bones of Orpheus, then the city of Libethra would be destroyed by a boar. The citizens paid little regard to the oracle, thinking that no other beast was big or mighty enough to take their city, while a boar was bold rather than powerful. But when it seemed good to the god the following events befell the citizens. About midday a shepherd was asleep leaning against the grave of Orpheus, and even as he slept he began to sing poetry of Orpheus in a loud and sweet voice. Those who were pasturing or tilling nearest to him left their several tasks and gathered together to hear the shepherd sing in his sleep. And jostling one another and striving who could get nearest the shepherd they overturned the pillar, the urn fell from it and broke, and the sun saw whatever was left of the bones of Orpheus. Immediately when night came the god sent heavy rain, and the river Sys [Boar], one of the torrents about Olympus, on this occasion threw down the walls of Libethra, overturning sanctuaries of gods and houses of men, and drowning the inhabitants and all the animals in the city. When Libethra was now a city of ruin, the Macedonians of Dium, according to my friend of Larisa, carried the bones of Orpheus to their own country. Whoever has devoted himself to the study of poetry knows that the hymns of Orpheus are all very short, and that the total number of them is not great. The Lycomidae know them and chant them over the ritual of the mysteries. For poetic beauty they may be said to come next to the hymns of Homer, while they have even been more honoured by the gods.[1]

The Thracian Bacchanals were responsible for killing Orpheus in a drug-induced frenzy.[2] These Dionysian Maenads who killed Orpheus created a new custom among the Thracian men, that is, one of fighting in battle drunk, the Latin poet Horace writes: "In goblets meant for gladness, the Thracians fight, in drink they battle, banish that custom here, away with it; defend our

[1] Pausanias, *Description of Greece* IX.30.6-12.

[2] Apollodorus, *The Library* I.3.2; Ovid, *Metamorphoses* XI.1-20.

modest Bacchus from barbarous blood-staining brawl."[1] Pausanias writes:

> Orpheus excelled his predecessors in the beauty of his verse, and reached a high degree of power because he was believed to have discovered mysteries, purification from sins, cures of diseases and means of averting divine wrath. But they say that the women of the Thracians plotted his death, because he persuaded their husbands to accompany him in his wanderings but dared not carry out their intention through fear of their husbands. Flushed with wine, however, they dared the deed, and hereafter the custom of their men has been to march to battle drunk.[2]

The Thracians had a reputation for their fine wines that were exported throughout the region.[3] The Thracians also had a reputation for heavy drinking, especially of unmixed wine, and the Thracian custom was to knock back the beverage in a single breath, which they called "tossing it off."[4] The Thracians also drank a potent barley wine called *pinon* and *bryton* (beer), which in his treatise *On Drunkenness*, Aristotle says *pinon* caused the drinker to fall backwards: "But a peculiar thing happens in the case of barley drinks, or the so-called *pinon.* Under the influence of all other intoxicants, those who get drunk fall in all directions, sometimes to the left, or to the right, or on their faces, or on their backs. But those who get drunk on *pinon* only fall backwards and lie supine."[5]

The Thracian Seuthes succeeded Sitalces as the king of the Odrysians, whose territory extended from Constantinople (modern Istanbul) along the seacoast of the Black Sea as far as the river Danube.[6] Athenaeus cites Xenophon, who attended a dinner party held by Seuthes:

> Xenophon, with customary elegance, describes in the *Anabasis* as occurring at the symposium held in the house of the Thracian Seuthes. He says: "When they had poured libations and sang the paean, the Thracians rose up to begin the programme, and danced in armour to a flute accompaniment. They leaped high and lightly, and

[1] Horace, *Carmina* I.27.
[2] Pausanias, *Description of Greece* IX.30.4-5.
[3] Xenophon, *Symposium* IV.41; Athenaeus, *Deipnosophistae* I.31.
[4] Athenaeus, *Deipnosophistae* XI.781; XII.534.
[5] Ibid. X.447.
[6] Thucydides, *History of the Peloponnesian War* II.96.1-3.

> brandished their knives. At the climax one struck the other; and all the audience thought that he had received a deadly blow. Down he fell with artful grace, and all the Paphlagonians at the dinner shouted aloud. Then the first dancer despoiled the other of his arms and made his exit with the Sitalces song, while other Thracians carried off the victim as though he were dead. But he wasn't hurt at all... They poured libations at the conclusion of dinner and offered them to Hermes, not, as in later times, to Zeus the Fulfiller. For Hermes is regarded as the patron of sleep. So they pour the libation to him also when the tongues of the animals are cut out on leaving the dinner. Tongues are sacred to him because he is the god of eloquence."[1]

According to Plato, the Thracian men and women would drink and pour wine over themselves:

> So let us deal more fully with the subject of drunkenness pure and simple, and the question is—ought we to deal with it as the Scythians and Persians do and the Carthaginians also, and Celts, Iberians and the Thracians, who are all warlike races, or as you Spartans do; for you, as you say, abstain from it all together, whereas the Scythians and Thracians, both men and women, take their neat wine and let it pour down over their clothes, and regard this practice of theirs as a noble and splendid practice...[2]

According to Athenaeus, this practice of excessive drinking extended to virtually all the Thracians:

> And Theompompus in the twenty-second book, when giving an account of the people of Chalcidice, in Thrace says: "For it happened that they despised the noblest pursuits, and were pretty much devoted to drinking, laziness, and excessive license." But the fact is that all Thracians are deep drinkers. Hence Callimachus, also, said: "For verily he loathed drinking wine greedily in a long Thracian draught, but was content with a small bowl." In the fifteenth book Theompompus says of the people of Methymna: "They ate their daily food in a sumptuous manner, reclining and drinking, but they did no deed in keeping with their lavish expenditures. And so

[1] Athenaeus, *Deipnosophistae* I.15-16.
[2] Plato, *Laws* 637D.

> the tyrant Cleomis stopped them from these practices; he was the one who tied up in sacks the procuresses who were in the habit of seducing well-born women of the free class, as well as three or four of the most conspicuous harlots, and ordered them to be drowned in the sea." Hermippus, too, in his work *On the Seven Sages*, says that Periander did the same. And Theompompus in the second book of his *History of Philip,* says that "the Illyrians dine and drink seated, and even bring their wives to parties; and it is good form for the women to pledge any of the guests, no matter who they may be. They conduct their husbands home from drinking-bouts...The people of Ardia (he continues) own 300,000 bondmen who are like helots. They get drunk every day and have parties, and are too uncontrolled in their predilection for eating and drinking."[1]

Orpheus, according to legend an Odrysian by birth, was credited with inventing the Mysteries of Dionysus.[2] According to the Roman philosopher Macrobius, the Thracian priests also consumed wine in their rituals: "the Ligureans in Thrace have a shrine dedicated to Liber (Dionysus) from which oracles are given. In this shrine the soothsayers drink large draughts of wine before delivering their prophecies..."[3] The Thracians had an ancient reputation for being a clever people who were careful in religious matters.[4] Besides Dionysus, the Thracians worshipped Ares and Artemis, the latter the daughter of Demeter, while their princes worshipped Hermes.[5] Herodotus writes:

> But these Satrae, as far as our knowledge goes, have never been subject to any man; they alone of all Thracians have ever been and are to this day free; for they dwell on high mountains covered with forests of all kinds and snow; and they are warriors of high excellence. It is they who possess the place of divination sacred to Dionysus, which place is among the highest of their mountains; the Bessi, a clan of the Satrae, are the prophets of the shrine, and it is a priestess that utters the oracle, as at Delphi, nor is aught more of

[1] Athenaeus, *Deipnosophistae* X.442-443.
[2] Apollodorus, *The Library* I.3.2.
[3] Macrobius, *Saturnalia* I.18.
[4] Pausanias, *Description of Greece* IX.29.3.
[5] Herodotus, *Histories* V.7; Pausanias, *Description of Greece* VIII.37.6.

mystery here than there.[1]

In Thrace the Sun was identified with Father Liber, that is, Dionysus, and in his *Bacchae*, Euripides attributes the power of prophecy to this deity: "A prophet is this god: the Bacchic frenzy and ecstasy are full-fraught with prophecy: For, in his fullness when he floods our frame, he makes his maddened votaries tell the future."[2] Ausonius of Bordeaux writes: "The sons of (the Greek) Ogyges call me Bacchus, Egyptians think me Osiris, (the Thracian) Mysians name me Phanaces, Indians regard me as Dionysus, Roman rites make me Liber, the Arab race thinks me Adoneus, Lucianiacus the Universal God. I am Osiris of the Egyptians, Phanaces of the Mysians, Bacchus among the living, Adoneus among the dead, Fire-born, Twy-horned, Titan-slayer, Dionysus."[3] Macrobius identifies Dionysus with Apollo, who was also worshipped by the Thracians:

> Certainly Aristotle, writing in his *Inquiries into the Nature of the Divine*, states that Apollo and Liber Pater [Dionysus] are one and the same god, and among the many proofs of this statement he says also that the Ligureans in Thrace have a shrine dedicated to Liber from which oracles are given. In this shrine the soothsayers drink large draughts of wine before delivering their prophecies, just as in the temple of Apollo at Claros water is drunk before the oracles are pronounced.[4]

The Greeks adopted many customs from the Thracians named Pelasgians, who settled Greece and established their religion in the culture. Included in this were the famous Samothracian rites of the Cabeiri, which originated in Egypt and were established on the Thracian island of Samos by the Arcadians (Pelasgians), mystery rites of which none would speak.[5] Herodotus writes in his *Histories*:

> These customs then and others besides, which I shall show, were taken by the Greeks from the Egyptians. It was not so with the ithyphallic images of Hermes, the making of these came from the Pelasgians, from whom the Athenians were the first of all Greeks

[1] Herodotus, *Histories* VII.111.
[2] Macrobius, *Saturnalia* I.18.
[3] Ausonius of Bordeaux, *Epigrams* 48-49.
[4] Macrobius, *Saturnalia* I.18.
[5] Dionysius of Halicarnassus, *Roman Antiquities* I.68.

> to take it, and then handed it on to others. For the Athenians were then already counted as Greeks when the Pelasgians came to dwell in the land with them, and thereby began to be considered as Greeks. Whoever has been initiated into the rites of the Cabeiri, which the Samothracians learnt from the Pelasgians who came to dwell among the Athenians, and it is from them that the Samothracians take their rites. The Athenians, then, were the first Greeks to make ithyphallic images of Hermes, and this they did because the Pelasgians taught them. The Pelasgians told a certain sacred tale about this, which is set forth in the Samothracian mysteries.[1]

There was a time when the Pelasgians used no names for their deities, but then adopted names from the Egyptians, and passed these on to the Greeks. Herodotus recounts this history:

> Formerly, in their sacrifices, the Pelasgians called upon gods (this I know, for I was told at Dodona) without giving name or appellation to any; for they had not as yet heard of such. They called them gods…Then, after a long while, they learned the names first of the rest of the gods, which came to them from Egypt, and, much later, the name of Dionysus; and presently they inquired of the oracle at Dodona concerning the names; for this place of divination is held to be the most ancient in Hellas, and at that time it was the only one. When the Pelasgians, then, inquired at Dodona if they should adopt the names that had come from foreign parts, the oracle bade them use the names. From that time onwards they used the names of the gods in their sacrifices; and the Greeks received these later from the Pelasgians.[2]

Herodotus also writes: "The fashions of divination in Thebes in Egypt and Dodona are like to one another; moreover the practice of divining from the sacrificed victim has also come from Egypt. It would seem too that the Egyptians were the first people to establish solemn assemblies, and processions, and services, the Greeks learnt all this from them."[3] Porphyry writes, "Chæremon the Stoic, therefore, in his narration of the Egyptian

[1] Herodotus, *Histories* II.51.
[2] Ibid. II.52.
[3] Ibid. II.57.

priests, who, he says, were considered by the Egyptians as philosophers, informs us, that they chose temples, as the places in which they might philosophize."[1] According to Herodotus, the Greeks also adopted the doctrine of reincarnation from the Egyptians:

> It is believed in Egypt that the rulers of the lower world are Demeter [Isis] and Dionysus [Osiris]. Moreover, the Egyptians were the first to teach that the human soul is immortal, and at the death of the body enters into some other living thing then coming to birth; and after passing through all creatures of land, sea, and air (which cycle it completes in three thousand years) it enters once more into a human body at birth. Some of the Greeks, early and late, have used this doctrine as if it were their own...[2]

The Thracians worshipped the goddess Artemis, the daughter of Demeter, and the Greeks called her Bubustis after Bubastis, the "city of Pasht" on the Egyptian Delta, where the inhabitants celebrated the festivals of Artemis, who was worshipped as the cat-headed Pasht.[3] Herodotus writes:

> When the people are on their way to Bubastis they go by river, men and women together, a great number of each in every boat. Some of the women make a noise with rattles, others play flutes all the way, while the rest of the women, and the men, sing and clap their hands. As they journey by river to Bubastis whenever they come near any other town they bring their boat near the bank; then some of the women do as I have said, while some shout mockery of the women of the town, others dance, and others stand up and expose their persons. This they do whenever they come beside any riverside town. But when they have reached Bubastis, they make a festival with great sacrifices, and more wine is drunk at this feast than in the whole year beside. Men and women (but not children) are wont to assemble there to the number of seven hundred thousand, as the people of the place say.[4]

Pythagoras (*ca.* 6th century B.C.) was another famous Thracian who followed the tradition attached to the name of Orpheus and his son Musaeus,

[1] Porphyry, *Abstinence from Animal Food* IV.6.
[2] Herodotus, *Histories* II.123.
[3] Ibid. II.59; II.137; Pausanias, *Description of Greece* VIII.37.6.
[4] Herodotus, *Histories* II.60.

while Socrates and his student Plato followed Pythagoras, and Plato's student Aristotle, also a Thracian by birth, ultimately rebelled against this tradition. Pythagoras was born on the island of Samos off the Thracian coast, also called Samothrace, translating into Greek as "sacred island."[1] One of the sayings of Pythagoras was "Follow God," and his disciples were so greatly admired that they were called "prophets who declare the voice of God."[2]

According to numerous authorities, Pythagoras traveled to Egypt to learn from Sonchis, highest prophet of the Egyptians, and he also traveled throughout Asia and Europe, gaining knowledge from the Chal'deans, the Magi, the Brahmans, the Gauls (Druids), the Tyrrhenians, and the Syrian Pherecydes.[3] The Pythagoreans wore their hair long and abstained from sex, although Pythagoras was said to have been in love with Theanô of Croton, the first woman philosopher and writer of poetry.[4] The Thracian philosopher Proclus writes:

> Pythagoras was from the island of Samos. His birth is said to have been forecast by prophecy, and he is said to have been born of a virgin—"prophecy" is a heavenly communication from God—; it is said that he kept company with Thales, another of the Seven Sages, that he went to Egypt and the River Nile, associated with the wise men, and learned geometry and the principles of prophecy; that he went also to Babylon, where he consulted the astrologers and the Magi—the meaning of "Magism," according to Plato, is the veneration of God—; it is said that his wisdom came to be such, that wise men of all countries came to him and that he attained so advance a stage of philosophy as to repudiate wealth and perform miracles.[5]

Pythagoras traveled abroad to learn from other priesthoods, and bringing his knowledge back to Europe, he settled at Croton in Italy where he and his followers successfully governed the population. Diogenes writes:

---

[1] Diodorus Siculus, *The Library of History* I.69.4; III.55.8-9; V.47.1-2.

[2] Boëthius, *The Consolation of Philosophy* I.140; Diogenes Laertius, *Lives of Eminent Philosophers* VIII.4.

[3] Diogenes Laertius, *Lives of Eminent Philosophers* VIII.1-2; Clement of Alexandria, *Stromata* I.62-71.

[4] Athenaeus, *Deipnosophistae* IV.163; XIII.599; Clement of Alexandria, *Stromata* I.80; III.24.

[5] Proclus, *Commentary on the Pythagorean Golden Verses* 91.

> While still young, so eager was [Pythagoras] for knowledge, he left his own country and had himself initiated into all the mysteries and rites not only of Greece but also of foreign countries...he learnt the Egyptian language, so we learn from Antiphon in his book *On Men of Outstanding Merit*, and he also journeyed among the Chaldaeans and Magi. Then while in Crete he went down into the cave of Ida with Epimenides; he also entered the Egyptian sanctuaries, and was told their secret lore concerning the gods. After that he returned to Samos to find his country under the tyranny of Polycrates; so he sailed away to Croton in Italy, and there he laid down a constitution for the Italian Greeks, and he and his followers were held in great estimation; for, being nearly three hundred in number, so well did they govern the state that its constitution was in effect a true aristocracy [government by the best].[1]

Like Orpheus and Socrates, Pythagoras made enemies with the locals and was murdered (497 B.C.), after which his followers settled in Metapontum and lasted there for a very long time.[2] The Roman historian Justin writes:

> [Pythagoras] was born on Samos, the son of Maratus [Mnemarchus], a wealthy merchant, and after making great progress in a philosophical education there, he had set out first for Egypt and then Babylonia to learn astronomy and examine the questions of the origins of the universe, ultimately attaining the heights of learning. On his return he had gone to Crete and Sparta to familiarize himself with the legal systems of Minos and Lycurgus, which were famous at that time. Endowed with all this learning he came to Croton where he succeeded by his personal authority in reconverting to the practice of frugality a population which had lapsed into luxurious living. He made a point of praising virtue every day, and of listing the ills of profligacy and the disasters which had struck states ruined by this malaise; and he rallied the common people to such an enthusiastic espousal of the simple life that it came to appear incredible that any of them had ever been profligate. It was a frequent practice of him to give instructions to married women apart from their husbands,

[1] Diogenes Laertius, *Lives of Eminent Philosophers* VIII.2-3.

[2] Clement of Alexandria, *Stromata* I.62.

> and to children apart from their parents. To the former he would teach chastity and obedience to their husbands, to the latter decorum and literary studies, and he continually impressed upon all of them the importance of temperance as the parent of the virtues. And by reiterating his arguments he achieved with success that the married women cast off their gold-embroidered robes and the other finery of their rank as being the promoters of extravagance, took them to the temple of Juno and consecrated them to the goddess herself, thus declaring publicly that the real finery of married women was chastity, not clothes. How far he succeeded with the young people can clearly be gauged from his triumph over the recalcitrant female temper.[1]

Clement writes, "Ion of Chios in his *Triads* records that Pythagoras attributed some of his work to Orpheus."[2] Among the mysteries in which Pythagoras was initiated included the sacred rites at Libethra in Thrace, which is near Mount Olympus and Pieria, said to be the burial place of Orpheus. According to the Greek philosopher Iamblichus, Pythagoras was a follower of Orpheus and the Orphics:

> [A] clear model for Pythagorean theology according to number is found in [the writings of] Orpheus. It is certainly no longer doubtful that Pythagoras took his inspiration from Orpheus when he organized his treatise *On Gods*, which he also entitled *The Sacred Discourse*, since it sprang from the most mystic part of the Orphic corpus…It is certainly clear from this *Sacred Discourse* who gave Pythagoras the discourse on gods, for it says: "This [discourse] is what I Pythagoras, son of Mnemarchus, learned in initiation in the Thracian Libethra from Aglaophamus the initiator"…from this it is clear that he derived the idea of the essence of the gods as defined by numbers from the Orphics.[3]

According to Iamblichus, Pythagoras was initiated into the Eleusinian Mysteries among other mystery rites:

> In general, they say Pythagoras was a zealous admirer of

[1] Justin, *Epitome of the Philippic History of Pompeius Trogus* XX.4.3-13.
[2] Clement of Alexandria, *Stromata* I.131.
[3] Iamblichus, *De Vita Pythagorica* 145-147.

> Orpheus' style and rhetorical art…He proclaimed their purification rites and what are called "mystic initiations," and he had most accurate knowledge of these things. Moreover, they say that he made a synthesis of divine philosophy and worship of the gods, having learned some things from the Orphics, others from the Egyptian priests, some from the Chaldeans and the Magi, then from the mystic rites in Eleusis, Imbros, Samothrace, Lemnos, and whatever was to be learned from mystic associations; and some from the Celts [Druids] and Iberians.[1]

According to Clement, Pythagoras consorted with the prophets of Egypt, "who induced him to undergo circumcision in order to enter their shrines and learn from the Egyptians their mystic philosophy. He was found with the best of the Chaldaeans and the Magi…"[2] According to Iamblichus, both Pythagoras and Plato were perpetuating the Hermetic teachings in their philosophy.[3] The Latin philosopher Seneca states; "Pythagoras declares that our souls experience a change when we enter a temple and behold the images of the gods face to face, and await the utterances of an oracle."[4] Clement writes, "the great Pythagoras applied himself ceaselessly to acquiring knowledge of the future: so did Abaris the Hyperborean…Zoroaster the Mede…Socrates of Athens…Philochorus in volume one of his work *On Divination* records the fact that Orpheus was a seer."[5]

Thales of Miletus (*ca.* 638–546 B.C.) was the earliest Greek natural philosopher, and he also studied philosophy in Egypt, and was later named among the venerable Seven Sages.[6] Orpheus and Thales published their doctrines and discourses in the form of poems, in contrast to Pythagoras and Socrates who never wrote down anything.[7] The Greek philosopher Plutarch writes: "Witness to this also are the wisest of the Greeks: Solon, Thales, Plato, Eudoxus, Pythagoras, who came to Egypt and consorted with the priests; and in this number some would include Lycurgus also…Pythagoras, as it seems, was greatly admired, and he also greatly admired the Egyptian priests, and

1 Ibid. 151.
2 Clement of Alexandria, *Stromata* I.66.
3 Iamblichus, *Mysteries of the Egyptians, Chaldeans, and Assyrians* I.2.
4 Seneca, *Epistulae Morales* XCIV.43.
5 Clement of Alexandria, *Stromata* I.133-134.
6 Plutarch, *Moralia* 875.
7 Ibid. 328; 402.

copying their symbolism and occult teachings, incorporated his doctrines in enigmas."[1]

The Jewish scholar Josephus states, "Once more, the first Greek philosophers to treat celestial and divine subjects such as Pherecydes of Syros, Pythagoras and Thales, were, as the world unanimously admits, in their scant productions the disciples of the Egyptians and Chaldaeans."[2] Speaking of Pythagoras, he adds: "In practicing and repeating these precepts he (Pythagoras) was imitating and appropriating the doctrines of Jews and Thracians. In fact, it is actually said that great man introduced many points of Jewish law into his philosophy."[3] Josephus later reveals that the great philosophers of ancient Thrace and Greece were monotheists: "Pythagoras, Anaxagoras, Plato, the Stoics who succeeded him, and indeed nearly all the philosophers appear to have held similar views concerning the nature of God. These, however, addressed their philosophy to the few, and did not venture to divulge their true beliefs to the masses who had their own preconceived opinions…"[4]

As a result of his travels to Egypt and other lands, Pythagoras gained wisdom and a reputation that surpassed all others, Isocrates writes:

> If one were not determined to make haste, one might cite many admirable instances of the piety of the Egyptians, that piety which I am neither the first nor the only one to have observed; on the contrary, many contemporaries and predecessors have remarked of it, of whom Pythagoras of Samos is one. On a visit to Egypt he became a student of the religion of the people, and was the first to bring to the Greeks all philosophy, and more conspicuously than others he seriously interested himself in sacrifices and in ceremonial purity, since he believed that even if he should gain thereby no greater reward from the gods, among men, at any rate, his reputation would be greatly enhanced. And this indeed happened to him. For so greatly did he surpass all others in reputation that all the younger men desired to be his pupils, and their elders were more pleased to see their sons staying in his company than attending to their

[1] Ibid. 354.
[2] Josephus, *Against Apion* I.14.
[3] Ibid. I.165.
[4] Ibid. II.168.

> private affairs. And these reports we cannot disbelieve; for even now persons who profess to be followers of his teachings are more admired when silent than are those who have the greatest renown for eloquence.[1]

Diogenes writes, "[Pythagoras] divides man's life into four quarters thus: 'twenty years a boy, twenty years a youth, twenty years a young man, twenty years an old man; and these four periods correspond to the four seasons, the boy to spring, the youth to summer, the young man to autumn, and the old man to winter,' meaning by youth one not yet grown up and by a young man a man of mature age."[2] Clement adds, "Heraclides of Pontus records that Pythagoras taught that happiness is the scientific knowledge of the perfection of the numbers of the soul."[3] The *Pythagorean Golden Verses* contains the following advice: "But whatever pains mortals suffer through the divine workings of fate, whatever lot you have, bear it and do not be angry."[4]

In what appears to be a drug, Plutarch says in his *Isis* that "juniper" was mixed with other ingredients while reading the sacred writings, and this was used by Pythagoras "as a charm and a cure for the emotional and irrational of the soul."[5] Pythagoras promoted the doctrine of the immortality of the soul, the Greek historian Diodorus writes: "Pythagoras of Samos and some others of the ancient philosophers declared that the souls of men are immortal, and also that, in accordance with this doctrine, souls foreknow the future at the moment in death when they are departing from the bodies."[6] According to Athenaeus, although they believed in immortality, the Pythagoreans did not believe in suicide:

> Euxitheus the Pythagorean, as the Peripatetic Clearchus tells us in the second book of his *Lives*, was wont to say that the souls of all beings are imprisoned in the body and in this hither life as a punishment, and that the god has ordained that if they refuse to abide in these until he of his own will releases them, they will then be plunged in more and greater torments. Wherefore all persons,

[1] Isocrates, *Busiris* 28-30.
[2] Diogenes Laertius, *Lives of Eminent Philosophers* VIII.10.
[3] Clement of Alexandria, *Stromata* II.130.
[4] *Pythagorean Golden Verses* 17.
[5] Plutarch, *Isis and Osiris* 80.
[6] Diodorus Siculus, *The Library of History* XVIII.1.1.

> dreading the violence of the higher powers, are afraid to depart from this life of their own motion, and gladly welcome only the death which comes in old age, being persuaded that this release of their souls comes with the approval of the higher powers.[1]

Plutarch says that the followers of Pythagoras were exponents of the moderate use of wine, and were told, "Do not walk under the vine."

> For what reason was it forbidden the priests of Jupiter to touch ivy or to pass along a road overhung by a vine growing on a tree? Is this second question like the precepts: "Do not eat seated on a stool," "Do not sit on a peck measure," "Do not step over a broom"? For the followers of Pythagoras did not really fear these things nor guard against them, but forbade other things through these. Likewise the walking under a vine had reference to wine, signifying that it is not right for the priest to get drunk; for wine is over the heads of drunken men, and they are oppressed and humbled thereby, when they should be above it and always master this pleasure, not be mastered by it.[2]

The Pythagoreans were encouraged to follow their own path, and not the herd, the Jewish philosopher Philo writes:

> Now we are told that the saintly company of the Pythagoreans teaches among other excellent doctrines this also, "walk not on the highways." This does not mean that we should climb steep hills—the school was not prescribing foot-weariness—but it indicates by this figure that in our words and deeds we should not follow popular and beaten tracks…Rising above the opinions of the common herd they have opened up a new pathway, in which the outside world can never tread, for studying and discerning truths, and have brought to light the ideal forms which none of the unclean may touch.[3]

Referring to his early teacher Sotion, a Pythagorean philosopher, the Latin philosopher Seneca explains why Pythagoras and others refrained from consuming animal flesh:

> Sotion used to tell me why Pythagoras abstained from animal

[1] Athenaeus, *Deipnosophistae* IV.157.
[2] Plutarch, *Moralia* 290.
[3] Philo, *Every Good Man is Free* I.1-3.

> food, and why, in later times, Sextius did also. In each case, the reason was different, but it was in each case a noble reason. Sextius believed that man had enough sustenance without resorting to blood, and that a habit of cruelty is formed whenever butchery is practised for pleasure...Pythagoras, on the other hand, held that all beings were inter-related, and that there was a system of exchange between souls which transmigrated from one bodily shape into another. If one may believe him, no soul perished or ceases from its functions at all, except for a tiny interval—when it is being poured from one body into another.[1]

According to Solon among others, Orpheus also abstained from meat.[2] According to Plato, the abstinence from animal flesh was part of the "Orphic life:"

> The custom of men sacrificing one another is, in fact, one that survives even now among many peoples; whereas amongst others we hear of how the opposite custom existed, when they were forbidden so much as to eat an ox, and their offerings to the gods consisted, not of animals, but of cakes of meal and grain steeped in honey, and other such bloodless sacrifices, and from flesh they abstained as though it were unholy to eat it or to stain with blood the altars of the gods; instead of that, those of us men who then existed lived what is called an "Orphic life," keeping wholly to inanimate food and, contrariwise, abstained wholly from things animate.[3]

According to Ammianus, Pythagoras as well as Socrates had a guardian spirit:

> For the theologians maintain that there are associated with all men at their birth, but without interference with the established course of destiny, certain divinities of that sort, as directors of their conduct; but they are visible only to a very few, whom their many merits have raised to eminence. And this oracles and writers of distinction have shown; among the latter is also the comic poet Menander, in whom we read these two senarii: "A daemon is assigned to every man

[1] Seneca, *Epistulae Morales* CVIII.17-19.
[2] Plutarch, *Moralia* 159; Euripides, *Hippolytus* 953.
[3] Plato, *Laws* 782C.

> at birth, to be the leader of his life." Likewise from the immortal poems of Homer we are given to understand that it was not the gods of heaven that spoke with brave men, and stood by them or aided them as they fought, but that guardian spirits attended them; and it was through reliance upon their special support that Pythagoras, Socrates, and Numa Pompilius became famous...[1]

The Latin philosopher Cicero says that Pythagoras, Plato's Old Academy, Aristotle's Peripatetics, and the Stoics who followed Zeno all embraced and practiced divination:

> And since they thought that the human mind, when in an irrational and unconscious state, and moving by its own free and untrammeled impulse, was inspired in two ways, the one by frenzy and the other by dreams...In this same category also were the frenzied prophecies of soothsayers and seers...Now my opinion is that, in sanctioning such usages, the ancients were influenced more by actual results than convinced by reason. However certain very subtle arguments to prove the trustworthiness of divination have been gathered by philosophers...for example, Socrates and all of the Socratic School and Zeno and his followers, continued in the faith of the ancient philosophers and in agreement with the Old Academy and with the Peripatetics. Their predecessor, Pythagoras, who even wished to be considered an auger himself, gave the weight of his great name to the same practice...The Stoics, on the other hand...defended nearly every sort of divination.[2]

Cicero writes, "And poetic inspiration also proves that there is a divine power with the human soul. Democritus says that no one can be a great poet without being in a state of frenzy, and Plato says the same thing."[3] Cicero then asks, "what ground is there to doubt the absolute truth of my position? For I have on my side reason, facts, peoples, and races, both Greek and barbarian, our own ancestors, the unvarying belief of all ages, the greatest philosophers, the poets, the wisest men, the builders of cities, and the founders of republics," to which he adds, "the truth is that no other argument of any sort is advanced

[1] Ammianus Marcellinus, *The Chronicles of Events* XXI.14.3-5.
[2] Cicero, *De Divinatione* I.2-3.
[3] Ibid. I.37.80.

to show the futility of the various kinds of divination which I have mentioned except the fact that it is difficult to give the cause or reason of every kind of divination."[1] Quoting Socrates, Plato defends divination in his *Phaedrus*:

> "For if it were a simple fact that insanity is an evil, the saying would be true; but in reality the greatest of blessings come to us through madness, when it is sent as a gift of the gods. For the prophetess at Delphi and the priestesses at Dodona when they have been mad have conferred many splendid benefits upon Greece both in private and public affairs, but few or none when they have been in their right minds; and if we should speak of the Sibyl and all the others who by prophetic inspiration have foretold many things to many persons and thereby made them fortunate afterwards, anyone can see that we should speak a long time. And it is worth while to adduce also the fact that those men of old who invented names thought that madness was neither shameful nor disgraceful; otherwise they would not have connected the very word mania with the noblest of arts, that which foretells the future, by calling it the manic art. No, they gave this name thinking that mania, when it comes by gifts of the gods, is a noble thing...And a third kind of possession and madness comes from the Muses. This takes hold upon a gentle and pure soul, arouses it and inspires it to songs and other poetry, and thus by adorning countless deeds of the ancients educates later generations. But he who without the divine madness comes to the doors of the Muses, confident that he will be a good poet by art, meets with no success, and the poetry of the sane man vanishes into nothingness before that of the inspired madman. All these noble results of inspired madness I can mention, and many more. Therefore let us not be afraid on that point, and let no one disturb and frighten us by saying that the reasonable friend should be preferred by him who is in a frenzy. Let him show in addition that love is not sent from heaven for the advantage of love and beloved alike, and we will grant him the prize of victory. We, on our part, must prove that such madness is given by the gods for our greatest happiness; and our proof will not be believed by the merely clever,

[1] Ibid. I.39.84-85.

> but will be accepted by the truly wise. First then, we must learn the truth about the soul divine and human by observing how it acts and is acted upon. And the beginning of our proof is as follows: Every soul is immortal."[1]

According to Xenophon, Socrates was "so religious that he did nothing without counsel from the gods," later adding, "When anyone was in need of help that human wisdom was unable to give he advised him to resort to divination…"[2] A Pythian oracle once pronounced Socrates "the wisest of mankind."[3] In his *Metaphysics*, Aristotle writes, "There are two innovations which may fairly be ascribed to Socrates: inductive reasoning and general definition. Both of these are associated with the starting-point of scientific knowledge."[4] According to Diogenes, Socrates "would extol leisure as the best of possessions…There is, he said, only one good, that is knowledge, and only one evil, that is, ignorance, wealth and good birth bring their possessor no dignity, but on the contrary evil…Moreover in his old age he learnt to play the lyre…He used to say that his supernatural sign warned him beforehand of the future…"[5]

Socrates believed in an all-knowing and omnipotent deity (i.e. God), who was not aloof from human affairs.[6] Like Plato, Xenophon was a student of Socrates, and quotes his teacher in his *Memorabilia*: "'Yet again, in so far as we are powerless of ourselves to foresee what is expedient for the future, the gods lend us their aid, revealing the issues of divination to inquirers, and teaching them how to obtain the best results.' 'With you Socrates, they seem to deal even more friendly than with other men, if it is true that, even unasked, they warn you by signs what to do and what not to do.'"[7]

Xenophon reflects on the death of his teacher Socrates:

> I have often wondered by what arguments those who drew up the indictment against Socrates could persuade the Athenians that his life was forfeit to the state. The indictment against him was to this effect: *Socrates is guilty of rejecting the gods acknowledged by the*

[1] Plato, *Phaedrus* 244A-245C.
[2] Xenophon, *Memorabilia* IV.7.10; IV.8.11.
[3] Pliny, *Natural History* VIII.118-120.
[4] Aristotle, *Metaphysics* 107B.
[5] Diogenes Laertius, *Lives of Eminent Philosophers* II.31-32.
[6] Xenophon, *Memorabilia* I.4.13-18.
[7] Ibid. IV.3.12.

> *state and bringing in strange deities: he is also guilty of corrupting the youth.* First then, that he rejected the gods acknowledged by the state—what evidence did they produce of that? He offered sacrifices constantly and made no secret of it, now in his home, now at the altars of the state temples, and he made use of divination with as little secrecy. Indeed it had become notorious that Socrates claimed to be guided by "the deity"; it was out of this claim, I think, that the charge of bringing in strange deities arose. He was no more bringing in anything strange than are other believers in divination, who rely on augury, oracles, coincidences and sacrifices. For these men's belief is not that the birds or the folk they meet by accident know what profits the inquirer, but that they are the instruments by which the gods make this known; and that was Socrates belief too. Only, whereas most men say that the birds or the folk they meet dissuade or encourage them, Socrates said what he meant: for he said that the deity gave him a sign...For, like most men, indeed, he believed that the gods are heedful of mankind, but with an important difference; for whereas they do not believe in the omniscience of the gods, Socrates thought that they know all things, our words and deeds and secret purposes; that they are present everywhere, and grant signs to men of all that concerns man. I wonder, then, how the Athenians can have been persuaded that Socrates was a freethinker, when he never said or did anything contrary to sound religion, and his utterances about the gods and his behaviour towards them were the words and actions of a man who is truly religious and deserves to be thought so.[1]

According to Diogenes, Plato went to Italy to learn from the Pythagorean philosophers, after which he traveled to Egypt.[2] The Palestinian Christian Eusebius cites the book *On the Good* by the Pythagorean philosopher Numenius that connects the teachings of Pythagoras with those of Plato: "But when one has spoken upon thus point, and sealed it by the testimonies of Plato, it will be necessary to go back and connect it with the precepts of Pythagoras, and to appeal to the other nations of good repute, bringing

[1] Ibid. I.1.1-19.
[2] Diogenes Laertius, *Lives of Eminent Philosophers* III.6.

forward their rites and doctrines, and their institutions which are formed in agreement with those of Plato, all that the Brahmans, and Jews, and Magi, and Egyptians arranged."[1]

In *The revolt of the Academics against Plato,* Numenius states the following in the fragments preserved by Eusebius: "But Plato had been a Pythagorean, and knew that Socrates for the same reason took much sayings from no other source than that, and had known what he was saying; and so he too wrapped up his subjects in a manner that was neither usual nor plain to understand..."[2] According to Plutarch, "Socrates went about seeking to solve the questions of what arguments Pythagoras used to carry conviction," and that even among their students, "Plato and Socrates did not win over many."[3]

The Christian Hippolytus writes, "Plato, being asked by some one, 'What is philosophy?' replied, 'It is a separation of the soul from the body'...For these suppose that there is a transition of souls from one body to another, as also Empedocles, adopting the principles of Pythagoras, affirms."[4] Plato's student Aristotle was acquainted with the teachings of Pythagoras, evidenced by his monograph *On the Pythagoreans* and the numerous references to Pythagorean doctrine through his *Metaphysics*. In *Metaphysics*, Aristotle writes:

> In his youth Plato first became acquainted with Cratylus and the Heraclitean doctrines—that the whole sensible world is always in a state of flux, and that there is no scientific knowledge of it—and in after years he still held these opinions. And when Socrates, disregarding the physical universe and confining his study to moral questions, sought in this sphere for the universal and was the first to concentrate upon definition, Plato followed him...[5]

Clement writes, "Plato does not deny importing from abroad the best parts into his philosophy, and admits a visit to Egypt...Plato's continual respect for non-Greeks is clearly revealed: he recalls that he, like Pythagoras, learned the majority of his finest theories among foreigners."[6] Plato's teachings

[1] Eusebius, *Praeparatio Evangelica* IX.7.
[2] Ibid. XIV.5.
[3] Plutarch *Moralia* 328; 516.
[4] Hippolytus, *The Refutation of all Heresies* 21.
[5] Aristotle, *Metaphysics* 987a-b.
[6] Clement of Alexandria, *Stromata* I.66-68.

concerning the immortality of the soul and the judgments in Hades and things of this nature were also taught by the priests in India,[1] to which Megesthenes adds, "All that has been said regarding nature by the ancients is asserted also by the philosophers outside of Greece, on the one part in India by the Brahmans, and on the other in Syria by the people called the Jews."[2] Clement writes:

> Plato in the *Cratylus* attributes to Orpheus the doctrine that the soul is in the body as a punishment. Here are his words: "Some people say that it is the burial place of the soul, which is at the present time entombed in it. Because the soul uses the body to mention whatever it would mention, the body is rightly called the soul's burial place. However, it is the followers of Orpheus who seem to have established the name above all others, saying the soul is paying the penalty for acts that have earned the penalty." It is worth noting Philolaus' remark. The follower of Pythagoras says, "The theologians and seers of old are witnesses that the soul is yoked to the body to undergo acts of punishment and is buried in it as in a grave."[3]

Clement cites the work *To Philometor* by the Alexandrian Jew Aristobulus (2nd century B.C.): "'Plato too was a follower of our system of law, and it is obvious that he had spent a lot of time on each of its precepts...it is clear enough that the philosopher I mentioned, a man of wide learning, took a great deal from it, much as Pythagoras borrowed a great deal from us to form his own philosophic system.' The Pythagorean philosopher Numenius wrote directly: 'What is Plato but Moses speaking Greek?'"[4] Directly quoting Aristobulus, Eusebius writes:

> It is evident that Plato closely followed our legislation, and has carefully studied the several precepts contained in it...as also Pythagoras transferred many of our precepts and inserted them in his own system of doctrines...Now it seems to me that he [God] has been very carefully followed in all by Pythagoras, and Socrates, and Plato, who said that they heard the voice of God, when they were

[1] Strabo, *Geographica* XV.1.59; Megesthenes, *Fragment* 41.
[2] Megesthenes, *Fragment* 42.
[3] Clement of Alexandria, *Stromata* III.16-17.
[4] Ibid. I.150.

> contemplating the arrangement of the universe so accurately made and indissolubly combined by God. Moreover, Orpheus, in verses taken from his writings in the *Sacred Legend*, thus sets forth the doctrine that all things are governed by divine power, and that they have had a beginning, and that God is over all.[1]

It is evident from the *Orphic Fragments* that Orpheus embraced monotheism, that states: "One power, one God, one vast and flaming heav'n, one Universal frame..."[2] According to Herodotus, the ritual called Orphic and Bacchic "is in truth Egyptian and Pythagorean..."[3] Plutarch provides some insight into the orientation of the followers of Orpheus: "Plato banters the followers of Orpheus for declaring that for those who have lived rightly, there is laid up in Hades a treasure of everlasting intoxication."[4] Clement says that among others, Pythagoras, Plato, and the Thracian Democritus, the latter a leading ethical exponent of happiness, acquired their greatest knowledge from foreign sources:

> Pythagoras is recorded as the disciple of Sonchis, highest prophet of the Egyptians, Plato of Sechnuphis of Heliopolis, Eudoxus the Cnidian of Chonuphis, another Egyptian...Again he [Plato] shows us his knowledge of prophecy when he introduces the character who proclaims Lachesis' dictum to the souls casting lots, so as to predict the future. In the *Timaeus* he introduces Solon, in all his wisdom, learning from the foreigner...Democritus appropriated the ethical teachings of the Babylonians. He is said to have included an interpretation of the stele of Acicarus in his own writings, as can be recognized from his writing; "These are Democritus' words." Yes, and actually boasting of his wide learning, he says somewhere, "I traveled over more of the earth than any human beings of my day, made the most extensive researches, viewed more climates and countries, listened to more intellectuals. No one has ever surpassed me in the composition of geometrical figures accompanied by demonstration, not even the Egyptians called Arpedonaptae...I spent in all eighty years abroad." He traveled to Babylon, Persia,

[1] Eusebius, *Praeparation Evangelica* XIII.12.
[2] *Orphic Fragments* IV.16.
[3] Herodotus, *Histories* II.81.
[4] Plutarch, *Lucullus* I.2.

> and Egypt and studied with Magi and priests. Pythagoras was enthusiastic about Zoroaster, the Persian Magus...Alexander, in his work *On Pythagorean Symbols*, records that Pythagoras was a pupil of the Assyrian Zaratus, and claims in addition that Pythagoras learned from the Gauls [Druids] and the Brahmans.[1]

According to Diodorus, in their sacred books the Egyptian priests recorded the names of the visitors they received from Thrace and Greece, who all returned home possessing knowledge that made them greatly admired:

> But now that we have examined these matters, we must enumerate what Greeks, who have won fame for their wisdom and learning, visited Egypt in ancient times, in order to become acquainted with its customs and learning. For the priests of Egypt recount from the records of their sacred books that they were visited in early times by Orpheus, Musaeus, Melampus, and Daedalus, also by the poet Homer and Lycurgus of Sparta, later by Solon of Athens and the philosopher Plato, and that there also came Pythagoras of Samos and the mathematician Eudoxus, as well as Democritus of Abdera and Oenopides of Chios. As evidence for the visits of all these men they point in some cases to their statues and in others to places or buildings which bear their names, and they offer proofs from the branch of learning which each one of the men pursued, arguing that all the things for which they were admired among the Greeks were transferred from Egypt.[2]

Diodorus later elaborates on the influence of Egyptian culture upon the Thracian and Greek philosophers and their societies:

> Lycurgus also and Plato and Solon, they say, incorporated many Egyptian customs into their own legislation. And Pythagoras learned from the Egyptians his teachings about the gods, his geometrical propositions and theory of numbers, as well as the transmigration of the soul into every living thing. Democritus also, as they assert, spent five years among them and was instructed in many matters relating to astrology. Oenopides likewise passed some time with the priests and astrologers...Like the others, Eudoxus studied astrology

[1] Clement of Alexandria, *Stromata* I.69-70.
[2] Diodorus Siculus, *The Library of History* I.96.1-3.

> with them and acquired a notable fame for the great amount of useful knowledge he disseminated among the Greeks.[1]

Clement links the philosophy and religion of the Greeks to the Indo-European priesthood, namely to the prophets of the Egyptians, the Assyrian Chaldaeans, the Persian Magi, the Indian Brahmans, and the British Druids among others:

> So philosophy reached a climax long ago among non-Greeks as something precious, and shone brightly through the peoples and later reached the Greeks. Its main authorities were the prophets of the Egyptians, the Chaldaeans among the Assyrians, the Druids among the Gauls, the Samanaeans in Bactria...the Magi among the Persians (who by their magic powers actually foretold the Savior's birth, arriving in the land of Judaea led by a star), the Gymnosophists in India, and other non-Greek philosophers.[2]

Plutarch writes, "Plutarch of Chaeronea in Boeotia informs us that the rites of the Dionysia and of the Panathenaic festival, and indeed those of the Thesmorphia and of the Eleusinian mysteries, were imported into Attica by Orpheus, an Odrysian, and that after visiting Egypt he transplanted the ritual of Isis and Osiris into the ceremonies of Deo and Dionysus."[3] According to Diodorus, "Orpheus also made many changes in the practices and for that reason the rites which had by established by Dionysus were also called 'Orphic.'"[4] Diodorus also states that Musaeus, the son of Orpheus, was in charge of the initiatory rites at Eleusis.[5]

The Greeks adopted the Orphic rites and the Eleusinian mysteries from the Thracians, the latter of which, as legend has it, was taught to the Thracian Eumolpus by the Goddess Demeter.[6] The Thracians settled Athens under Eumolpus, conducting the Eleusinian rites until the Visigoth king Alaric destroyed the temple the fourth century A.D.[7] Isocrates records the history of the invasion of Athens by the Thracians led my Eumolpus: "For our country was invaded by the Thracians, led by Eumolpus, son of Poseidon, who

[1] Ibid. I.98.1-3.
[2] Clement of Alexandria, *Stromata* I.71.
[3] Plutarch, *Fragment* 212.
[4] Diodorus Siculus, *The Library of History* III.65.
[5] Ibid. IV.25.
[6] *Homeric Hymn to Demeter* 474-476; Strabo, *Geographica* X.3.1.
[7] Pausanias, *Description of Greece* I.38.2-3.

disputed the possession of Athens with Erechtheus, alleging that Poseidon had appropriated the city before Athena…"[1] According to Pausanias, the war concluded when the Eleusinians, or Thracians, agreed with the Athenians over power-sharing, giving the Eleusinians control over the mystery rites, and the Athenians control over everything else: "This Eumolpus, they say came from Thrace…When the Eleusinians fought with the Athenians, Erechtheus, king of the Athenians, was killed, as was Immaradus, son of Eumolpus. These were the terms on which they concluded the war: the Eleusinians were to have independent control over the mysteries, but all things else were to be subject to the Athenians."[2]

According to Pausanias, the mystic rites of Demeter were sometimes performed in circuits of large unhewn stone:[3]

> The people of Pheneüs have also a sanctuary of Demeter, surnamed Eleusinian, and they perform a ritual to the goddess, saying that the ceremonies at Eleusis are the same as those established among themselves. For Naüs, they assert, came to them because of an oracle from Delphi, being a grandson of Eumolpus. Beside the sanctuary of the Eleusinian has been set up Petroma, as it is called, consisting of two large stones fitted one to the other. When every other year they celebrate what they call the Greater Rites, they open these stones. They take from out them writings that refer to the rites, read them in the hearing of the initiated and return them on the same night. Most Phenaetians, too, I know, take an oath by the Petroma in the most important affairs.[4]

According to Pausanias, at Corinth in Greece, "there are three temples close together, one of Apollo, one of Artemis, and the third of Dionysus…Of the gods the Aeginetans worship most Hecate, in whose honour every year they celebrate mystic rites which, they say, Orpheus the Thracian established among them."[5] Hecate was identified with Aphrodite, and considered the daughter of Demeter she traveled between the two worlds, and was the "third"

[1] Isocrates, *Panathenaicus* 193.
[2] Pausanias, *Description of Greece* I.38.2-3.
[3] Ibid. II.34.10.
[4] Ibid. VIII.15.1-2.
[5] Ibid. II.30.2.

that joined the Mother Goddess Demeter with her the daughter Persephone.[1] The Greeks identified Demeter with the Egyptian Isis, who was identified with Ceres, who in later tradition as Circe was treated as the daughter of Hecate.[2] In his *Metamorphoses*, the Roman satirist Apuleius reveals the identity of "Isis of a thousand names:" "The Phrygians, first born of men, call me Mother of the Gods, goddess of Pessinus; the inhabitants of Attica, Minerva of Cecrops City (Athens), the Cypriots being amid the seas, Venus of Paphos, the arrow bearing Cretans, Diana of Dictynna; the triple tongued Sicilians, Proserpina of the Styx; the original Elyeusian, Ceres of Attica, some (call me) Juno, others Bellona, some Hecate, others the one from Rhamnus."[3] According to Pliny, Venus is called Juno, Isis, and the Mother of the Gods, and she creates a dew that produced drugs with the potency "equal to that of the nectar of the gods."[4]

Demosthenes writes: "You must magnify the Goddess of Order who loves what is right and preserves every city and every land…each [man] must reflect that he is being watched by hallowed and inexorable Justice, who, as Orpheus, that prophet of our most sacred mysteries, tells us, sits beside the throne of Zeus and oversees all the works of men. Each must keep watch and ward lest he shame that goddess."[5] Pausanias declares that the Savior Maid Artemis was the goddess worshipped by the Pelasgian Thracians: "The Great Goddesses are Demeter and the Maid…And the Maid is called Saviour by the Arcadians (Pelasgians)."[6] According to Pausanias, the temple of the Savior Maid in Greece was built either by Orpheus, or Abaris the Hyperborean, and of the latter more will be said.[7] According to Diodorus, "The initiation rite, which is celebrated by the Athenians at Eleusis, the most famous, one may venture, of them all, and that of Samothrace, and the one practiced in Thrace among the Cicones, whence Orpheus came who introduced them—those are all handed down in the form of a mystery (i.e., secretly)…" Pausanias writes:

[1] Ibid. II.30.2; Euripides, *Ion* 1048; Apuleius, *Apologia* 31; Virgil, *Aenid* VI.147-160; *Paris Magic Papyrus* 1403; 2257.

[2] Herodotus, *Histories* II.59; Euripides, *Helen* 1301-68; Apollodorus, *Bibliotheca* II.1.3; Diodorus Siculus, *The Library of History* I.13.4; I.96.5; Augustine, *De Civitate Dei* VII.16; VIII.27.

[3] Apuleius, *Metamorphoses* XI.5.

[4] Pliny, *Natural History* II.38.

[5] Demosthenes, *Against Aristocrates* I.11.

[6] Pausanias, *Description of Greece* VIII.31.1.

[7] Ibid. III.13.2.

> Above the temple of Athena is a grove, surrounded by a wall, of Artemis named Saviour, by whom they swear their most solemn oaths. No man may enter the grove except the priests. These priests are natives, chosen chiefly because of their high birth. Opposite the grove of the Saviour is a sanctuary of Dionysus surnamed Torch. In his honour they celebrate a festival called the Feast of Torches, when they bring by night firebrands into the sanctuary, and set up bowls of wine throughout the whole city.[1]

Athenaeus states that the followers of Pythagoras were devoted to playing the pipe-flute and three-stringed lyre called a cithara, the archetype of the guitar.[2] Clement claims that the Phrygian named Agnis created the three-stringed lyre.[3] Some attributed the invention of musical rhythms to the Phrygian Dactyls of Ida who worshipped the Great Goddess Rhea Cybelê, and these Phrygian Dactyls were said to have brought the music of the flute to Greece, the invention of the traverse flute being attributed to Satyrus of Phrygia.[4] Athenaeus says that music, especially music played in the Phrygian mode, was believed to cure disease.[5]

According to Athenaeus the ancients believed that music could cure violence, intemperance, gloominess, and says that the Pythagorean Cleinias "would always take his lyre and play on it whenever it happened that he was exasperated to the point of anger."[6] According to Plutarch, the Pythagoreans employed the lyre before sleeping "as a charm and a cure for the emotion and irrational in the soul."[7] Athenaeus says that the Thracian Getae conducted their diplomatic negotiations will playing the cithara, and the stringed instrument call the phoenix was employed by Thracian princes at their banquets to promote harmony and good-will.[8]

There were three melodic scales: Phrygian, Lydian and Dorian.[9] According to Athenaeus, the Phrygian and Lydian modes were taught to the

[1] Ibid. VII.27.3.
[2] Athenaeus, *Deipnosophistae* IV.184; XIV.624.
[3] Clement of Alexandria, *Stromata* I.75-76.
[4] Ibid. I.73; I.76; Plutarch, *Moralia* 1132.
[5] Athenaeus, *Deipnosophistae* XIV.624.
[6] Ibid. XIV.623-627.
[7] Plutarch, *Moralia* 384.
[8] Athenaeus, *Deipnosophistae* XIV.627-637.
[9] Ibid. XIV.635.

Greeks by the Phrygians and Lydians of Asia minor who immigrated to the Peloponnesus with Pelops: "Hence also Telestes of Selinus says 'The first to sing the Phrygian strains in honour of the Mountain Mother, amid the flutes beside the mixing-bowls of the Greeks, were they who came in the company of Pelops; and the Greeks struck up the Lydian hymn with the high-pitched twang of the lyre.'" Thamyris of Thrace was a legendary blind bard, a renowned singer and composer, and he was credited with inventing the Dorian mode.[1]

Pausanias describes a Greek painting that depicts Orpheus playing music in the grove of Persephone:

> Turning our gaze again to the lower part of the picture we see... Orpheus sitting on what seems to be a sort of hill; he grasps with his left hand a harp, and with his right he touches a willow. It is the branches that he touches, and he is leaning against the tree. The grove seems to be that of Persephone, where grow, as Homer thought, black poplars and willows. The appearance of Orpheus is Greek, and neither his garb nor his head-gear is Thracian...after him is Pelias, sitting on a chair, with grey hair and grey beard, and looking at Orpheus...Thamyris is sitting near Pelias. He has lost the sight of his eyes; his attitude is one of utter dejection; his hair and beard are long; at his feet lies thrown a lyre with its horns and string broken. Above him is Marsyas, sitting on a rock, and by his side is Olympus, with the appearance of a boy in the bloom of his youth learning to play the flute. The Phrygians in Celaenae hold that the river passing through the city was once this great flute-player, and they also hold that the Song of the Mother, an air for the flute, was composed by Marsyas. They say too that they repelled the army of the Gauls by the aid of Marsyas, who defended them against the barbarians by the water from the river and the music of his flute.[2]

In his *Discourses*, the Greek writer Dio Chrysostom associates music with intoxication:

> [W]hile other people are moved to song and dance by drink, with

[1] Homer, *Iliad* II.594-600; Plutarch, *Moralia* 1132; Clement of Alexandria, *Stromata* I.76.

[2] Pausanias, *Description of Greece* X.30.6-9.

you the opposite is true—song is the occasion of drunkenness and frenzy. So while wine's natural effect is as we have seen, producing inability to preserve one's self-control, but on the contrary forcing those who use it stupidly and in excess to commit many distasteful acts, yet men intoxicated by song are in far worse condition than those who are crazed by wine; and what is more, at the very start and not by easy stages as at a drinking party, such men, I say, are to be found nowhere but in Alexandria. Among certain barbarians, it is true, we are told that a mild kind of intoxication is produced by the fumes of a certain incense when burned. After inhaling it they are joyful and get up and laugh, and behave in all respects like men who have been drinking, and yet without doing injury to one another; but of the Greeks you alone reach that state through ears and voice, and you talk more foolishly than do those barbarians, and you stagger worse and are more like men suffering the after-effects of a debauch. And yet the arts of the Muses and Apollo are kindly gifts and pleasing. For Apollo is addressed as Healer and as Averter-of-Evil, in the belief that he turns men aside from misfortune and implants health in soul and body, not sickness or madness; and the Muses are called maidens, implying their modesty and their chastity. Furthermore, music is believed to have been invented by men for the healing of their emotions, and especially for transforming souls which are in a harsh and savage state. That is why even some philosophers attune themselves to the lyre at dawn, thereby striving to quell the confusion caused by their dreams. And it is with song that we sacrifice to the gods, for the purpose of insuring order and stability in ourselves. And there is, moreover, a different type of song, accompanied by the flute, that is employed at the time of mourning, as men attempt, no doubt, to heal the harshness and the relentlessness of their grief and to mitigate the pain by means of song, song that operates scarce noticed amid lament, just as physicians, by bathing and softening wounds that are inflamed, remove the pain. And the spell of music has been deemed no less appropriate also in social gatherings, because it brings harmony and order spontaneously into the soul along with a kindred influence abates the unsteadiness that

> comes from delight in wine—I mean the very influence blended with which the unsteadiness itself is brought into tune and tempered to moderation. All this, of course, in the present instance has been reversed and changed to its opposite. For it is not by the Muses but by a kind of Corybantes that you are possessed, and you lend credibility to the mythologizings of the poets, since they do indeed bring upon the scene creatures called Bacchants, who have been maddened by song, and Satyrs too.[1]

Regino of Prüm (9th century A.D.), a German priest and musician, recognized the power of music, and relates a story about Pythagoras:

> It should be known, moreover, that the characters of men are recognized through music. The lustful and impudent person is delighted by the lascivious modes or, hearing them frequently, is softened and made effeminate. On the contrary, the more severe and more arrogant disposition either rejoices in the harsher modes or is stirred up by the harsher ones...For people who are of the harsher type are delighted by the harsher modes, while those of the softer type are pacified by the milder ones. Such is the disposition of characters that all are governed and ruled by songs, so that both for proceeding into battle or for retreating use is made of the trumpet which excites or, again, subdues the courage of the soul. Song gives sleep and takes it away, and even casts out cares and anxieties or brings them back. It stirs up anger, it counsels kindness, and it also cures diseases of the body. It is, indeed, common knowledge how often song has repressed irascible minds; how many wonders have been performed on the affliction of bodies and minds. Finally, according to Cicero, when youths drunk with wine, lured by the song of the tibia [aulos], broke down a virtuous woman's door so that they might violate her chastity, Pythagoras learned it was by the sound of the Phrygian mode that the souls of the youths had been incited to lust (for at that hour he was observing the course of the stars) and immediately, running to her, he warned the tibia player to change the mode and play a spondee [long slow notes]. And when she had done this, by playing the modes with slowness and dignity,

[1] Dio Chrysostom, *Discourses* XXXII.55-58.

the fury of the youths came to rest.[1]

Plutarch quotes Theophrastus who held that music has three sources: sorrow, joy, and religious ecstasy.[2] The cithara and the flute played the "libation-music" that attended the celebration of Dionysus.[3] The dithyramb was the dance music that was associated with the Dionysian cult and their rituals. The Greek poet Archilochus writes, "And I know how to lead off the sprightly dance of the Lord Dionysus, the dithyramb. I do it thunderstruck with wine."[4] According to Plutarch, music was introduced to counteract the effects of wine, adding that Aristoxenus says, "music was introduced forasmuch as wine makes the bodies and minds of those who overindulge in it disorderly, while music by its order and balance brings us to the opposite condition and soothes us. Hence Homer asserts that the ancients employed music as a remedy to meet this issue."[5] Athenaeus associates Dionysus with wine and drunkenness, and Apollo with order and tranquility:

> Philochorus says that the ancients, in pouring libations, do not always sing dithyrambs, but when the pour libations, they celebrate Dionysus with wine and drunkenness, but Apollo, in quiet and good order. Archilochus, at any rate, says: "For I know how to lead off, in my lovely song of lord Dionysus, the dithyramb, when my wits have been stricken with the thunder-bolt of wine." And Epicharmus, also, said in *Philoctetes*: "There can be no dithyramb when you drink water." It is plain, therefore, in light of what we have said, that music did not, at the beginning, make its way into feasts merely for the sake of shallow and ordinary pleasure, as some persons think.[6]

According to Herodotus, Arion was the greatest lyre-player of his age, and was the first man to compose and name the dithyramb.[7] Strabo relates the myth of Arion, and credits Terpander with the seven-stringed lyre:

> Arion, who, according to a myth told by Herodotus and his followers, safely escaped on a dolphin to Taenarum after being thrown into the sea by pirates. Now Arion played, and sang to, the

[1] Regino of Prüm, *De Harmonica Institutione* 6.
[2] Plutarch, *Moralia* 623.
[3] Athenaeus, *Deipnosophistae* XIV.638.
[4] Archilochus, *Fragment* 249.
[5] Plutarch, *Moralia* 1146.
[6] Athenaeus, *Deipnosophistae* XIV.628.
[7] Herodotus, *Histories* I.23.

> cithara; and Terpander, also, is said to have been an artist in the same music and to have been born in the same island, having been the first person to use the seven-stringed instead of the four-stringed lyre, as we are told in the verses attributed to him: "For thee I, having dismissed four-toned song, shall sing new hymns to the tune of the seven-stringed cithara."[1]

Terpander was influenced by the music of Orpheus, who Plutarch credits as having no musical predecessor to imitate:

> Terpander appears to have been eminent as an executant in singing to the cithara; thus it is recorded that he won four successive victories at the Pythian games...Alexander in his *Notices on Phrygia* said that Olympus first brought the music of the auloi to the Greeks, but that the Idaean Dactyls did so too; that Hyagnis was the first to play the auloi and that his son Marsyas came next, and after him Olympus; and that Terpander took as his models the hexameters of Homer and the music of Orpheus. But Orpheus evidently imitated no predecessor, as there were none as yet, unless it was composers of songs of the auloi, and Orpheus' work resembles theirs in no way.[2]

The Spartans possessed an extremely conservative attitude toward music and frowned on innovation, and according to Plutarch, "If anyone presumed to transgress in any way the rules of the good old music, they would not permit this; but even Terpander, one of the oldest and the best harp-player of his time as well as a devoted admirer of the deeds of the heroes, the Ephors none the less fined, and carried away his instrument and nailed it to a wall because he put in just one extra string for the sake of the variety in the notes; for they approved only the simpler melodies."[3]

According to Plutarch, the motions of the stars, the universe, and all things in existence were "said by Pythagoras, Archytas, Plato, and the rest of the ancient philosophers not to come into being or to be maintained without the influence of music; for they assert that God has shaped all things in a framework based on harmony."[4] Plutarch also reveals that Plato and Aristotle, and their best students, studied music. "Thus most of the Platonists

[1] Strabo, *Geographica* XIII.2.4.
[2] Plutarch, *Moralia* 1132.
[3] Ibid. 238.
[4] Ibid. 1147.

and the best of the Peripatetics have devoted their efforts to the composition of treatises on ancient music and its conception in their day; furthermore, the most learned grammarians and students of harmonies have also devoted much study to the subject."[1] Strabo writes:

> Plato, and even before his time the Pythagoreians, called philosophy music; and they say that the universe is constituted in accordance with harmony, assuming that every form of music is the work of the gods. And in this sense, also, the Muses are goddesses, and Apollo is leader of the Muses, and poetry as a whole is laudatory of the gods. And by the same course of reasoning they also attribute to music the upbuilding of morals, believing that everything which tends to correct the mind is close to the gods. Now most of the Greeks assigned to Dionysus, Apollo, Hecatê, the Muses, and above all to Demeter, everything of an orgiastic or Bacchic or choral nature, as well as the mystic element in initiation...But all educated men, and especially the musicians, are ministers of the Muses; and both these and those who have to do with divination are ministers of Apollo...[2]

The Latin rhetorician Quintilian (1st century A.D.) says that the ancient philosophers were simultaneously priests, poets, and musicians:

> Who is ignorant of the fact that music, of which I will speak first, was in ancient times the object not merely of intense study but of veneration: in fact Orpheus and Linus, to mention no others, were regarded as uniting the roles of musician, poet and philosopher. Both were of divine origin...So too Timagenes asserts that music is the oldest of the arts related to literature, a statement which is confirmed by the testimony of the greatest of poets in whose songs we read that the praise of heroes and of gods were sung to the music of the lyre at the feasts of kings...There can in any case be no doubt that some of those men whose wisdom is a household word have been earnest students of music: Pythagoras for instance and his followers popularised the belief, which they no doubt had received from earlier teachers, that the universe is constructed on the same

[1] Ibid. 1131.
[2] Strabo, *Geographica* X.3.10.

> principles which were afterwards imitated in the construction of the lyre…As for Plato, there are certain passages in his works, more especially in the *Timaeus*, which are quite unintelligible to those who have not studied the theory of music. But why speak only of the philosophers, whose master, Socrates, did not blush to receive instruction in playing the lyre even when far advanced in years. It is recorded that the greatest generals playing on the lyre and the pipe, and that the armies of Sparta were fired to martial ardour by the strains of music…It was not therefore without reason that Plato regarded the knowledge of music as necessary to his ideal statesman or politician, as he calls him; while the leaders even of that school, which in other respects is the strictest and most severe of all schools of philosophy [i.e., the Stoics], held that the wise man might well devote some of his attention to such studies. Lycurgus himself, the founder of the stern laws of Sparta, approved of the training supplied by music…there was actually a proverb among the Greeks, that the uneducated were far from the company of the Muses and Graces.[1]

Many Thracian customs were imported into Greece, including the Bendideian festival, held in honor of Goddess Bendis, the Thracian Mother of the Gods, this cult being adopted in Athens around 429 B.C. These mystical rites are mentioned by Plato in association with Socrates; apparently both philosophers were familiar with the cult. The "Berecyntian Goddess" of Phrygia was also the Mother of the Gods, whose priests were named Galli after the Phrygian river Gallus, and according to legend, those who drank from this river went mad.[2] Worshipping the Mother Goddess at Pessinus, "the Phrygians used to practise their orgiastic rites on the banks of the River Gallus, from which the eunuchs dedicated to the service of the goddess get their name."[3] According to the Greek poet Pindar, "One is the race of men, one is the race of gods, and from one mother do we both derive our breath…"[4] Strabo summarizes the influence of Thracian culture and music in ancient Greece:

> From its melody and rhythm and instruments, all Thracian

[1] Quintilian, *De Institutione Oratoria* I.10.9-22.
[2] Ovid, *Fasti* IV.355; Pliny, *Natural History* V.147.
[3] Herodian, *History of the Roman Empire* I.11.2; Appian, *Roman History* VII.56; etc.
[4] Pindar, *Nemean Odes* VI.1.

> music has been considered to be Asiatic. And this is clear, first, from the places where the Muses have been worshiped, for Pieria and Olympus and Pimpla and Leibethrum were in ancient times Thracian places and mountains, though they are now held by the Macedonians; and again, Helicon was consecrated to the Muses by the Thracians who settled in Boeotia [Greece], the same who consecrated the cave of the nymphs called Leibethrides. And again, those who devoted their attention to the music of earlier times are called Thracians, I mean Orpheus, Musaeus, and Thamyris; and Eumolpus ["sweet-singer"], too, got his name from there. And those writers who have consecrated the whole of Asia, as far as India, to Dionysus, derive the greater part of music from there. And one writer says, "striking the Asiatic cithara"; another calls flutes "Berecyntian" and "Phrygian"... [The Athenians] welcomed so many of the foreign rites that they were ridiculed therefore by comic writers; and among these were the Thracian and Phrygian rites. For instance, the Bendideian rites are mentioned by Plato, and the Phrygian by Demosthenes, when he casts the reproach upon Aeschines' mother and Aeschines himself that he was with her when she conducted initiations, that he joined her in leading the Dionysic march and that many a time he cried out "êvoe saboe," and "hyês attês, attês hyês"; for these words are in the ritual of Sabazius and the Mother.[1]

In Homer's *Odyssey*, Odysseus is presented twelve jars of the extremely potent unmixed Thracian wine by Maron, a priest of Apollo, "a drink for the gods" that was mixed twenty parts water to one part wine:

> With me I had a goat-skin of the dark, sweet wine, which Maron, son of Euanthes, had given me, the priest of Apollo, the god who used to watch over Ismarus. And he had given it me because we had protected him with his child and wife out of reverence; for he dwelt in a wooded grove of Phoebus Apollo. And he gave me splendid gifts: of well-wrought gold he gave me seven talents, and he gave me a mixing-bowl all of silver; and besides these, wine, wherewith he filled twelve jars in all, wine sweet and unmixed, a drink divine.

[1] Strabo, *Geographica* X.3.17-18.

> Not one of his slaves nor of the maids in his halls knew thereof, but himself and his dear wife, and one house-dame only. And as often as they drank that honey-sweet red wine he would fill one cup and pour it into twenty measures of water, and a smell would rise from the mixing-bowl marvelously sweet; then verily would one not choose to hold back.[1]

Athenaeus records that the Egyptian priests of Apollo consumed Dionysian wine in their rites:

> "In Naucratis," as Hermeias says in the second book *On the Gryneian Apollo,* "the people dine in the town hall [prytaneion] on the natal day of Hestia Prytantis and the festival of Dionysus, and again at the great gathering in honour of the Comaean Apollo, all appearing in white robes which even to this day they call their 'prytanic' clothes. After reclining they rise again, and kneeling, join in pouring a libation while the herald, acting as priest, recites the traditional prayers. After this they recline and all receive a pint of wine excepting the priests of Pythian Apollo and of Dionysus; for to each of these latter the wine is given in double quantity, as well as the portions of everything else...No woman may enter the town-hall except the flute-girl."[2]

According to Plutarch, among other writers, Aeschylus was said to have written his tragedies while drinking wine.[3] Horace believed that the poet benefited from wine, and declared sobriety belongs to the world of business, conducted in ancient times at the Forum and Libo's Well:

> If you follow old Cratinus, my learned Maecenas, no poems can please long, nor live, which are written by water-drinkers. From the moment Liber [Dionysus] enlisted brain-sick poets among his Satyrs and Fauns, the sweet Muses, as a rule, have had a scent of wine about them in the morning. Homer, by his praises of wine, is convicted as a winebibber. Even Father Ennius never sprang forth to tell of arms save after much drinking. "To the sober I shall assign the Forum and Libo's Well; the stern I shall debar from song." Ever

[1] Homer, *Odyssey* IX.195-210.
[2] Athenaeus, *Deipnosophistae* IV.149-150.
[3] Plutarch, *Moralia* 622.

> since I put forth this edict, poets have never ceased to vie in wine-drinking by night, and reek of it by day.[1]

It was customary of the Greeks to mix their wine with water to render it less potent, only drinking it unmixed after eating. Athenaeus writes:

> Philochorus has this: "Amphictyon, king of Athens, learned from Dionysus the art of mixing wine, and was the first to mix it. So it was that men came to stand upright, drinking wine mixed, whereas before they were bent double by the use of unmixed. Hence he founded an altar to the 'upright' Dionysus in the shrine of the Seasons; for these make ripe the fruit of the vine. Near it he also built an altar to the Nymphs to remind devotees of the mixing; for the Nymphs are said to be the nurses of Dionysus. He also instituted the custom of taking just a sip of unmixed wine after meat, as a proof of the power of the good god, but after that they might drink mixed wine, as much as each man chose. They were also to repeat over this cup the name of Zeus the Savior as a warning and reminder to drinkers that only when they drink in this fashion would they surely be safe." Plato in the second book of the *Laws* says that the use of wine is designed to promote health.[2]

Zeus the Savior, the Dionysian Father, was also the Good Daemon, whose supernatural qualities were identified with the wine, mixed by the Nymphs, the nurses of Dionysus. Athenaeus writes:

> Theophrastus in his work *On Drunkenness* says: "The unmixed wine which is given upon ending the dinner and which they call a 'toast in honour of the Good Daemon' is taken only in small quantity, just as a reminder, through a mere taste, of the strength in the god's generous gift; and they offer it after they have been satisfied with food, so that the amount drunk may be very small"... and Philochorus in the second book of his *Attic History* says: "In those days the custom was established that after the food only so much unmixed wine should be taken by all as should be a taste and ensample of the good god's power, but after that all other wine must be drunk mixed. Hence the Nymphs were called the nurses of

[1] Horace, *Epistles* XIX.1-12.

[2] Athenaeus, *Deipnosophistae* II.38.

Dionysus."[1]

Diodorus says that unmixed wine causes madness and stupor, but if used correctly the drinker could enjoy health benefits from its use:

> [T]he Boeotians and other Greeks and the Thracians...have established sacrifices every other year to Dionysus, and believe at that time the god reveals himself to human beings. Consequently in many Greek cities every other year Bacchic bands of women gather, and it is lawful for the maidens to carry the thrysus and to join in the frenzied revelry, crying out "Evuai!" honoring the god...And since the discovery of wine and the gift of it to human beings were the source of such great satisfaction to them, both because of the pleasure which derives from the drinking of it and because of the greater vigour which comes to the bodies of those who partake of it, it is the custom they say, when unmixed wine is served during a meal to greet it with the words, "To the Good Deity!" but when the cup is passed around after the meal diluted with water, to cry out, "To Zeus Savior!" For the drinking of unmixed wine results in a state of madness, but when it is mixed with the rain from Zeus the delight and pleasure continue, but the ill effects of madness and stupor is avoided.[2]

Plutarch compared wine to love, and reveals that water was consumed along with the mixed wine to mitigate against the harmful effects:

> With regards to wine we ought to talk as does Euripides with regard to Love: "Mayest thou be mine, but moderate be, I pray, yet ne'er abandon me." For wine is the most beneficial of beverages, the pleasantest of medicines, and the least cloying of appetizing things, provided that there is a happy combination of it with the occasion as well as with water. Water, not only the water that is mixed with the wine, but that which is drunk by itself in the interim between the draughts of the mixture, makes the mixture more innocent.[3]

The ancient Greeks praised the moderate use of wine, and were also keenly aware of the dangers of immoderation. Athenaeus writes:

[1] Ibid. XV.693.
[2] Diodorus Siculus, *The Library of History* IV.3.
[3] Plutarch, *Moralia* 132.

> And Panyasis says: "Wine is as great a boon to earthly creatures as fire. It is loyal, a defender from evil, a companion to solace every pain. Yea, wine is the desired portion of the feast and of merry-making, of the tripping dance and of yearning love. Therefore, thou shouldst receive and drink it at the feast with a glad heart, and when satisfied with food thou shouldst not sit still like a child, filled to over-flowing, oblivious of the mirth." And again: "But wine is the best gift of gods to men, sparkling wine; every song, every dance, every passionate love, goes with wine. It drives all sorrows from men's hearts when drunk in due measure, but when taken immoderately it is a bane."[1]

Athenaeus reports that the wine consumed during his time was capable of causing mass hallucinations: "A party of young fellows were drinking it, and became so wild when overheated by the liquor that they imagined they were sailing in a trireme, and that they were in a bad storm on the ocean. Finally they completely lost their senses, and tossed all the furniture and bedding out of the house as though upon the waters, convinced that the pilot directed them to lighten the ship because of the raging storm."[2]

According to Plutarch, Plato and Aesop found wine useful in examining the character of people: "Plato, too, holds that most men show their real natures most clearly when they drink...For wine reveals us and displays us by not allowing us to keep quiet; on the contrary, it destroys our artificial patterns of behaviour, taking us completely away from convention's tutorship, so to speak. Aesop and Plato, then—and any other in need of a method of examination,—find wine useful for this purpose..."[3] Athenaeus adds, "Philochorus says that drinkers not only reveal what they are, but also disclose the secrets of everybody else in their outspokenness. Hence the saying 'wine is truth also,' and 'wine revealeth the heart of man,'" adding, "And Antiphanes: 'One may hide all else, Phenidia, but not these two things—that he is drinking wine, and that he has fallen in love. Both of these betray him through his eyes and through his words, so that the more he denies, the more they make it plain.'"[4]

[1] Athenaeus, *Deipnosophistae* XV.693.
[2] Ibid. II.37.
[3] Plutarch, *Moralia* 645.
[4] Athenaeus, *Deipnosophistae* II.37.

In his *Dionysiaca,* the Egyptian poet Nonnos poetically describes a Dionysian wine party in which the revelers are driven to madness, except for Dionysus, who was given an amethyst by the Phrygian Goddess Rhea Cybelê that protected him from the madness induced by the potent wine:

> A band of Satyrs was with him: one stooped to gather the clusters, one received them into an empty vessel as they were cut… then Bacchos spread the fruitage in the pit he had dug…and filled it to the brim, he trod the grapes with dancing steps. The Satyrs also, shaking their hair madly in the wind, learnt from Dionysus how to do the like. They pulled tight the dappled skins of fawns over their shoulder, they shouted the song of Bacchos sounding tongue with tongue, crushing the fruit with many a skip of the foot, crying "Evoi!" The wine spurted up in the grapefilled hollow, the runlets were empurpled; pressed by the alternating tread the fruit bubbled out red juice with white foam. They scooped it up with oxhorns, instead of cups which had not yet been seen, so that ever after the cup of mixed wine took this divine name of Winehorn…And one when bubbling the windcharming drops of Bacchos as he turned his wobbling feet in zigzag jerks, crossing right over left in confusion as he wetted his hairy cheeks with Bacchos's drops. Another skipt up struck with a tippler's madness when he heard the horrid boom of the beating drumskin. One again who had drunk too deeply of the caredispelling wine purpled his dark beard with the rosy liquor. Another, turning his unsteady look towards a tree espied a Nymph half-hidden, unveiled, close at hand; and he would have crawled up the highest tree in the forest, feet slipping, hanging on by his toenails, had not Dionysos held him back. Near the fountains, another driven by the insane impulse of drunken excitement, chased a naked Naiad…To Dionysos alone had Rheia given the amethyst, which preserves the winedrinker from the tyranny of madness. Many of the horned Satyrs joined furiously in the festive dancing with sportive steps. One felt within him a new hot madness, the guide to love, and threw a hairy arm round a Bacchanal girl's waist. One shaken by the madness of mind-crazing drink laid hold of the girdle of a modest unwedded maid, and as she would have no love-making pulled her

> back by the dress and touched her rosy thighs from behind. Another dragged back a struggling mystic maiden while kindling the torch for the god's nightly dances, laid timid fingers upon her bosom and pressed the swelling circle of her firm breast. After the revels over his sweet fruit, Dionysos proudly entered the cave of the Cybeleïd goddess Rheia, waving bunches of grapes in his flowerloving hand, and taught Maionia the vigil of the feast.[1]

The mysterious Dactyli of Mount Ida were believed to have originated in Crete or Phrygia, and according to tradition they practiced the initiation rites that were adopted by the Samothracians and by Orpheus. According to Diodorus, these people were credited with the discovery of copper and iron, both revolutionary advances in technology, and of the institution of the Olympic games:

> The inhabitants of Crete claim that the oldest people of the island were those who were known as Eteocretans ["Genuine Cretans"], who were sprung from the soil itself…The first of these gods of whom tradition has left any record made their home in Crete about Mt. Idê and were called Idaean Dactyli…But some historians, and Ephorus is one of them, record that the Idaean Dactyli were in fact born on the Mt Idê which is in Phrygia and passed over to Europe together with Mygdon; and since they were wizards, they practised charms and initiatory rites and mysteries, and in the course of a sojourn in Samothrace they amazed the natives of that island not a little by their skill in such matters. And it was at this time, we are further told, that Orpheus, who was endowed with an exceptional gift of poesy and song, also became a pupil of theirs, and he was subsequently the first to introduce initiatory rites and mysteries to the Greeks. However this may be, the Idaean Dactyli of Crete, so tradition tells us, discovered both the use of fire and what the metals copper and iron are, as well as the means of working them, this being done in the territory of the city of Aptera at Berecynthus, as it is called; and since they were looked upon as the originators of this great blessing for the race of men, they were accorded immortal honors. And writers tell us that one of them was named Heracles,

[1] Nonnos, *Dionysiaca* XII.337-397.

> and excelled as he did in fame, he established the Olympic Games, and that the men of a later period thought, because the name was the same, that it was the son of Alcmenê who had founded the institution of the Olympic Games. And evidences for this, they tell us, are found in the fact that many women even to this day take their incantations from this god and make amulets in his name, on the ground that he was a wizard and practised the arts of initiatory rites; but they add that these things were indeed very far removed from the habits of the Heracles who was born to Alcmenê.[1]

Following the Egyptians, the Samothracians mysteries taught that the first gods are the Great Gods who were "Earth" and "Sky," and this was identical to the Great Mother Goddess and her consort Serapis (Osiris, Dionysus). The Latin scholar Varro writes:

> The first gods were *Caelum* "Sky" and *Terra* "Earth." These gods are the same as those who in Egypt are called Serapis and Isis... For Earth and Sky, as the mysteries of the Samothracians teach, are Great Gods, and these whom I have mentioned under many names are not those Great Gods whom Samothrace represents by two male statues of bronze which she has set up before the city-gates, nor are they, as the populace thinks, the Samothracian gods, who are really Castor and Pollux; but these are a male and a female, these are those whom the *Book of Augurs* mention in writing as "potent deities," for what the Samothracians call "powerful gods." These two, Sky and Earth, are a pair like life and body.[2]

The Corybantes were the musically inspired attendants of the Great Mother Goddess, the Phrygian Cybelê, who was worshipped in Phrygia and Crete. The Phrygian Corybantes, who were identified with the Idaean Dactyli, the Curetes, and the Cabeiri, performed the Samothracian mysteries. Strabo writes:

> [S]ome represent the Corybantes, the Cabeiri, the Idaean Dactyli, and the Telchines as identical with the Curetes...roughly speaking and in general, they represent them, one and all, as a kind of inspired people and as subject to Bacchic frenzy, and, in the

[1] Diodorus Siculus, *The Library of History* IV.64.
[2] Varro, *De Lingua Latina* V.57-59.

> guise of ministers, as inspiring terror at the celebration of the sacred rites by means of war-dances, accompanied by uproar and noise and cymbals and drums and arms, and also by flute and outcry; and consequently these rites are in a way regarded as having a common relationship, I mean these and those of the Samothracians and those in Lemnos and in several other places, because the divine ministers are called the same.[1]

Strabo associates these ministers called "Cabeiri" and "Corybantes" with the Great Mother Goddess, the Phrygian Cybelê and her consort Dionysus, and their practices with the Bendideian and Orphic rites of the Thracians:

> They invented names appropriate to the flute, and to the noises made by castanets, cymbals, and drums, and to the acclamations and shouts of "ev-ah," and stampings of the feet; and they also invented some of the names by which to designate the ministers, choral dancers, and attendants upon the sacred rites, I mean "Cabeiri" and Corybantes" and "Pans" and "Satyri" and "Tityri," and they called the god "Bacchus," and Rhea "Cybelê" or "Cybebê" or "Dindymene" according to the place where she was worshipped. Sabazius also belongs to the Phrygian group and in a way is the child of the Mother, since he too transmitted the rites of Dionysus. Also resembling these rites are the Cotytian and the Bendideian rites practiced among the Thracians, among whom the Orphic rites had their beginning...these rites resemble the Phrygian rites, and it is at least not unlikely that, just as the Phrygians themselves were colonists from Thrace, so also their sacred rites were borrowed from there.[2]

The Athenians obtained their religion and subsequent civilization from the Thracians, and these religious rites spread across the region. Pausanias writes: "Methapus was an Athenian by birth, and expert in the mysteries and founder of all kinds of rites. It was he who established the mysteries of the Cabeiri at Thebes..."[3] According to Herodotus, the temples of the Cabeiri were considered sacred ground, and only the priests were permitted to enter

[1] Strabo, *Geographica* X.3.7.
[2] Ibid. X.3.15-16.
[3] Pausanias, *Description of Greece* IV.1.7.

these sanctuaries.[1] Pausanias reveals that Demeter, the Eleusinian Goddess, was closely associated with the Cabeiri, and she was said to have given the Cabeiri a gift, along with knowledge of the mystery rites:

> [Y]ou come to a grove of Cabeirean Demeter and the Maid. The initiated are permitted to enter it. The sanctuary of the Cabeiri is some seven stades distant from this grove. I must ask the curious to forgive if I keep silence as to who the Cabeiri are, and what is the nature of the ritual performed in honor of them and the Mother. But there is nothing to prevent my declaring to all what the Thebans say was the origin of the ritual. They say that once there was in this place a city, with inhabitants called Cabeiri, and that Demeter came to know Prometheus, one of the Cabeiri, and Aethaeüs his son, and entrusted something to their keeping. What was entrusted to them, and what happened to it, seemed to me a sin to put into writing, but at any rate the rites are a gift of Demeter to the Cabeiri...The wrath of the Cabeiri no man may placate, as has been proved on many occasions...certain men of the army of Xerxes left behind with Mardonius in Boeotia entered the sanctuary of the Cabeiri, perhaps in the hope of great wealth, but rather, I suspect, to show their contempt of its gods; all these immediately were struck with madness, and flung themselves to their deaths into the sea or from the tops of precipices. Again, when Alexander after his victory wasted with fire all the Thebaïd, including Thebes itself, some men from Macedonia entered the sanctuary of the Cabeiri, as it was in enemy territory, and were destroyed by thunder and lightning from heaven.[2]

Nonnos associates the Bassarids with the Berecyntian Goddess when he writes: "The panspipes sounded a cheerheart melody...the Bassarid waved the Euian tambourines...The bold king heard the jubilation of the dance... the Berecyntian tune...he beheld the vine-god (Dionysus) near his porch..."[3] He later associates the Bassarids with the Cabeiri: "Companies of Bassarids marched to battle. One shaking the untidy clusters of her tresses to and fro,

[1] Herodotus, *Histories* III.37.
[2] Pausanias, *Description of Greece* IX.25.5-10.
[3] Nonnos, *Dionysiaca* XX.300-310.

armed herself with raging madness for battle with the waters, driven wildly along with restless dancing feet. One whose home was in the Samothracian cavern of the Cabeiroi, skipt about the peaks of Lebanon crooning the barbarous notes of Corybantian tune."[1] Nonnos writes, "The Dictaian Corybants joined battle...The Indian nation was ravaged by the steel of those mountaineer herdsmen, the Curetes...The Bassarid lifted her leafy weapon of war, and cast: from that Bacchos-hating generation many men's heads were brought low by the woman's thrysus."[2] Nonnos adds, "From Crete came grim warriors to join them, the Idaian Dactyloi, dwellers on a rock crag, earthborn Corybants..."[3] The Dactyli of Ida were identical to the Phrygian Curetes.[4] In his second book, the Latin historian Curtius declares that Phrygia "is also said to be the native land of the Idaean Dactyli, or Corybantes who, instructed by the Great Mother (Cybelê) first discovered the twofold use of iron, a most cruel tool of rage and not less useful as an aid to poverty and toil."[5]

Nonnos associates the Greek Euboeans with the Corybantes: "The Euboean battalions were ruled by shield-bearing Corybantes, guardians of Dionysus in his growing days: who in the Phrygian gulf beside mountainranging Rheia surrounded Bacchos still a child with their drumskin."[6] Nonnos also writes of "the Curetian tribe from the land of the Abantes," adding "the sacred soil of the Abantes, the earthborn stock of the ancient Curetes, whose life is the tune of pipes, whose life is the goodly noise of beaten swords, whose heart is set upon rhythmic circling of the feet and the shield-wise dancing."[7] Abantias was the old name for Euboea, and the Euboeans were called "Abantes."[8] The Abantes wore their hair long, and were experts in hand-to-hand fighting.[9] Strabo cites Aristotle of Chalcis who says that the Thracians who settled the Greek island called those who held it "Abantes," and Strabo identifies the "Chalcidians" as the Cabeiri.[10] The Latin historian Vellius writes, "The Athenians established

[1] Ibid. XLIII.307-314.
[2] Ibid. XXIX.215-225.
[3] Ibid. XIV.23-25.
[4] Lucretius, *De Rerum Natura* II.629; Pausanias, *Description of Greece* V.7.7.
[5] Quintus Curtius, *History of Alexander* II.
[6] Nonnos, *Dionysiaca* XIII.135-139.
[7] Ibid. XIII.155; XXXVI.280.
[8] Homer, *Iliad* II.536-542; Pliny, *Natural History* IV.64.
[9] Plutarch, *Theseus* V.1-3; Statius, *Thebaid* VII.370.
[10] Strabo, *Geographica* X.1.3; X.3.19.

colonies at Chalcis and Eretreia in Euboea...the Chalcidians, who...were of Attic origin, founded Cumae in Italy...At a considerable later period, a portion of the citizens of Cumae founded Naples."[1] According to Curtius, some of the Euboean tribes became degenerate after becoming ignorant of the traditions and native customs.[2]

According to Pausanias, the Apollo-worshipping Euboeans were also called Dryopes: "The men of Asine are the only members of the race of the Dryopes to pride themselves on the name to this day. The case is very different with the Euboeans of Styra. They too are Dryopes in origin...the people of Styra disdain the name of Dryopes...But the men of Asine take the greatest pleasure in being called Dryopes...For they have both a temple of Apollo and again a temple and ancient statue of Dryopes, whose mysteries they celebrate every year."[3] The Dryopes colonized the islands of the Peloponnesus from Mount Parnassus in Greece, where nearby stood the famous temple of Apollo at Delphi.[4]

According to Strabo and a number of other writers, the mystical Cabeiri were not only uniquely close to the gods, but they too were gods:

> [T]hey were called, not only ministers of gods, but also gods themselves...the author of *Phoronis* speaks of the Curetes as "flute-players" and "Phrygians"...Some call the Corybantes, and not the Curetes, "Phrygians," but the Curetes "Cretes," and say that the Curetes were the first people to don brazen armour in Euboea, and that on this account they were also called "Chalcidians"...some call the Corybantes sons of Cronus, but others say that the Corybantes were sons of Zeus and Calliopê and were identical with the Cabeiri, and that these went off to Samothrace, which in earlier times was called Melitê, and that their rites were mystical...they were called Cabeiri after mountain Cabeirus in Berecyntia. Some however, believe that the Curetes were the same as the Corybantes and were ministers of Hecatê...Pherecydes says that nine Corybantes were sprung from Apollo and Rhetia, and that they took up their abode in Samothrace; and that three Cabeiri and three nymphs were

[1] Vellius Paterculus, *Roman History* I.4.1-2.
[2] Quintus Curtius, *History of Alexander* IV.12.11.
[3] Pausanias, *Description of Greece* IV.34.11.
[4] Ibid. V.1.2.

> called Cabeirides and were the children of Cabeiro, the daughter of Proteus, and Hephaestus, and that their sacred rites were instated in honour of each triad. Now it has also happened that the Cabeiri are most honoured in Imbros and Lemnos, but they are also honoured in separate cities of the Troad; their names, however are kept secret... The Curetes and Corybantes were the same...Some call them natives of Ida, others settlers; but all agree that iron was first worked by these on Ida; and they use the term Phrygia for the Troad because, after Troy was sacked, the Phrygians, whose territory bordered on the Troad, got the mastery over it. And they suspect that both the Curetes and the Corybantes were offspring of the Idaean Dactyli; at any rate, the first hundred men born in Crete were called Idaean Dactyli, they say, and as offspring of these were born nine Curetes, and each of these begot ten children who were called Idaean Dactyli.[1]

Sacred to Dionysus and maddened by his inspiration, the women known as Bacchanals were associated with the Thracians by the Egyptian poet Tryphiodorus: "Not so doth the pleasant flute of Dionysus raging on the hills strike the Thracian women amid the thickets: who, smitten by the god, strains a wild eye and shakes her naked head dark-garlanded with ivy."[2] The mother of Alexander the Great, Olympias, was a Bacchanal and was also initiated into the mystery rites at Samothrace, Plutarch writes in his *Alexander*:

> [A]ll the women of these parts were addicted to the Orphic rites and the orgies of Dionysus from very ancient times...and imitated in many ways the practices of the Edonian women and the Thracian women about Mount Haemus, from whom, as it would seem, the word "threskeuein" [Thracian women] came to be applied to the celebration of extravagant and superstitious ceremonies. Now Olympias, who affected these divine possessions more zealously than other women, and carried out these divine inspirations in wilder fashion, used to provide the reveling companies with great tame serpents, which would often lift their heads from out the ivy and the mystic winnowing-baskets or coil themselves about the wands and

[1] Strabo, *Geographica* X.3.19-22.
[2] Tryphiodorus, *Ilios* 367-372.

garlands of the women, thus terrifying the men.[1]

The Dionysian and Orphic rites of the Thraco-Phrygians were intimately connected to the Mother of the Gods, who was thought to be a Dryad nymph, Plutarch writes in his *Caesar*:

> Now, the Romans have a goddess whom they call Bona, corresponding to the Greek Gynaeceia. The Phrygians claim this goddess as their own, and say that she was the mother of King Midas; the Romans say she was a Dryad nymph and the wife of Faunus; the Greeks that she was the unnameable one among the mothers of Dionysus. And this is the reason why the women cover their booths with vine-branches when they celebrate her festival, and why the sacred serpent is enthroned beside the goddess in conformity with the myth. It is not lawful for a man to attend the sacred ceremonies, nor even be in the same house when they are celebrated; but the women, apart by themselves, are said to perform many rites during their sacred service which are Orphic in their character.[2]

In his *Against Eutropius*, the Latin poet Claudian (4th century A.D.) tells of the fall of Phrygia to the Getic invaders:

> Such was Phrygia when the gods allowed it to be ravaged by Getic brigands. The barbarians burst in upon those cities so peaceful, so easy to capture. There was no hope of safety, no chance of escape. Long and peaceful ages had made the crumbling stones of her battlements to fall. Meanwhile Cybelê was seated amid the hallowed rocks of cold Ida, watching, as is her wont, the dance, and inciting the joyous Curetes to brandish their swords at the sound of the drum, when, lo, the golden-turreted crown, the eternal glory of her blessed hair, fell from off her head and, rolling from her brow, the castellated diadem is profaned in the dust. The Corybantes stopped in amazement at this omen. The mother of the gods wept; then spake thus in sorrow. "This is the portent that agèd Lachesis foretold long years ago. My fallen crown assures me that Phrygia's final crisis is upon her...Now fare thee well, land of Phrygia, farewell, walls

[1] Plutarch, *Alexander* II.5-6.
[2] Plutarch, *Caesar* IX.3.

doomed to the flames…"[1]

According to Justin, the original inhabitants of Macedonia were called Pelasgians, and King Midas ruled over them.[2] Justin also says that Macedonia was called Emathia in the former times, and was ruled by the Thraco-Phrygian inhabitants until Caranus and a large band of Greeks invaded from the south and supplanted Midas and the other rulers of the territory.[3] In the region of Macedonia stood Mount Bermium (Doxa), and according to Strabo this was the home of the Briges, who would later be known as Phrygian: "in earlier times it was occupied by the Briges, a tribe of Thracians; some of these crossed over into Asia and their name was changed to Phryges."[4] According to Herodotus, the famous rose gardens of King Midas were at the foot of Mount Bermium.[5] Strabo places the habitation of Midas in Phrygia (Asia Minor) by the Sangarius River.[6] According to Strabo, after the Trojan War, the Briges in Europe changed their name to Phryges, and crossed back into Asia Minor, slaying a ruler of Troy and gaining control of the Troad (Northwest Asia Minor).[7] Strabo says that the "Phrygians themselves are Brigians, a Thracian tribe."[8]

Arrian tells the story of the Gordian knot, which according to an oracle the one who would untie it would be lord over Asia:

> [Gordium], the acropolis, where was the palace of Gordius and his son Midas…Gordius' wagon and the knot of the chariot's yoke. There was a widespread tradition about the chariot around the countryside; Gordius, they said, was a poor man of the Phrygians of old, who tilled a scanty parcel…Once, as he was ploughing, an eagle settled on the yoke and stayed, perched there, till it was time to loose the oxen; Gordius was astonished at the portent, and went off to consult the Telmissian prophets, who were skilled in the interpretation of prodigies, inheriting—women and children too—the prophetic gift. Approaching a Telmissian village, he met a girl

[1] Claudian, *Against Eutropius* II.274-297.
[2] Justin, *Epitome of the Philippic History of Pompeius Trogus* VII.1.
[3] Ibid. VII.1-12.
[4] Strabo, *Geographica* [*Fragment*] VII.25.
[5] Herodotus, *Histories* VIII.138.
[6] Strabo, *Geographica* XII.5.3.
[7] Ibid. X.3.22; XII.8.3; XIV.5.29; *Fragment* VII.25.
[8] Ibid. VII.3.2.

> drawing water and told her the story of the eagle; she, being also of the prophetic line, bade him return to the spot and sacrifice to Zeus the King. So then Gordius begged her to come along with him and assist in the sacrifice; and at the spot duly sacrificed as she directed, married the girl, and had a son called Midas. Midas was already a grown man, handsome and noble, when the Phrygians were in trouble with civil war; they received an oracle that a chariot would bring them a king and he would stop the war. True enough, while they were discussing this, there arrived Midas, with his parents, and drove, chariot and all, into the assembly. The Phrygians, interpreting the oracle, decided he was the man whom the gods had told them would come in a chariot; and thereupon made him king, and he put an end to the civil war. The chariot of his father he set up in the acropolis as a thank-offering to Zeus the King for sending the eagle. Over and above this there was a story about the wagon, that anyone who should untie the knot of the yoke should be lord of Asia.[1]

A few generations prior to the Trojan War, Jason, the son of Cretheus, and the Argonauts, including Orpheus, sailed the ship Argo to the far side of the Black Sea to obtain the Golden Fleece. Before embarking upon this dangerous mission, Orpheus initiated Jason and the Argonauts into the Samothracian Mysteries.[2] Dispatched from Greece by Pelias, king of Thessaly, the hero and his companions succeeded in their endeavor with the help of Medea, the daughter of Aeetes, king of the Colchians. Medea had "learned all the powers which drugs possess," and Jason pledged to wed her if she helped him, and after capturing the fleece they fled the land pursued by Medea's brother Apsrytus, who was slain by Medea's magic. Jason returned to Corinth with Medea and gave her two children, and ten years later Creon, king of the Corinthians, offered Jason his daughter, which he accepted, incurring Medea's deadly wrath.[3]

According to Procopius, some writers have inaccurately "stated that the territory of the Trapezuntines is joined either by the Sani...or by the

[1] Arrian, *Anabasis of Alexander* II.3.1-8.

[2] Apollodorus, *The Library* I.3.2; Apollonius Rhodius, *Argonautica* I.20; Valerius Flaccus, *Argonautica* II.440; etc.

[3] Euripides, *Medea* 1ff; Valerius Flaccus, *Argonautica* II.611; Diodorus Siculus, *The Library of History* IV.45-56.

Colchians, calling another people Lazi, who are actually addressed by this name at the present day," adding, "it is impossible that the Lazi should not be the Colchians, because they inhabit the banks of the Phasis River; and the Colchians have merely changed their name at the present time to Lazi, just as nations of men and many other things do."[1] Procopius says that the Phasis River divides Asia from Europe and empties into the Euxine Sea (Black Sea), and the Lazi, or Colchians, inhabit the European side of the river.

Procopius and other writers believed that the Golden Fleece was not taken from the territory of the Colchians, but from the Trapezintines who occupied the Asian side of the Phasis River:

> [Pontus] ends at the land of the Colchians. And as one sails into it, the land on the right is inhabited by the Bithynians, and next after them by the Honoriate and the Paphlagonians…beyond them are the people called Pontici as far as the city of Trapezus and its boundaries. In that region are a number of towns on the coast, among which are Sinope and Amisus, and close of Amisus is the town called Themiscyra and the river Thermodon, where they say the army of the Amazons originated…From here the territory of the Trapezuntines extends to the village of Susurmena and the place called Rhizaeum, which is two days journey distant from Trapezus as one goes toward Lazica along the coast…This [Rhizaeum] was called Apsyrtus in ancient times, having come to be named after the man on account of his catastrophe. For in that place the natives say that Apsyrtus was removed from the world by the plot of Medea and Jason…It is now clear that one might with good reason wonder at those who assert that the Colchians are adjacent to the Trapezuntines. For on this hypothesis it would appear that after Jason in the company with Medea had captured the fleece, he actually did not flee toward Hellas and his own land, but backward to the Phasis river and the barbarians in the most remote interior… For I think that Jason would not have eluded Aeetes and got away from there with the fleece in the company with Medea, unless both the palace and other dwellings of the Colchians had been separated by the Phasis River from the place in which that fleece was lying;

[1] Procopius, *History of the Wars* VIII.1.8-10.

indeed the poets who have recorded the story imply that this was the case.[1]

According to Xenophon, Trapezus was "an inhabited Greek city on the Euxine Sea, a colony of Sinopians in the territory of Colchis."[2] Sinope was known for having the best red ochre.[3] According to Strabo, the best red ochre was produced in Cappadocia, and it was called "Sinopian."[4] Red ochre, or sinopus, was a drug used medically for eye pains and fluxes, eye tumors, vomiting, blood spitting, spleen disease, live disease, belly ailments, kidney disease, excessive menstruation, erysipelas (ergot poisoning), poisonous snake bites, and sea-serpent stings.[5] Procopius writes, "the honey which is produced in all the places around Trapezus is bitter, this being the only place where it is at variance with its established reputation."[6] According the Xenophon, "the soldiers who ate of the honey all went off their heads, and suffered from vomiting and diarrhea, and not one of them could stand up, but those who had eating a little were like people exceedingly drunk, while those who had eaten a great deal seemed like crazy, or even, in some cases, dying men."[7] Aristotle writes, "At Trapezus in Pontus honey from the boxwood has a heavy scent; and they say that healthy men go mad, but epileptics are cured by it immediately."[8]

The city of Mycenae was built in the late Minoan period (1580–1450 B.C.). According to Strabo, Perseus founded the city, and Sthenelus succeeded him, and they also ruled the city of Argos, adding that these two cities were in such close proximity that writers call the city at one time Mycenae, and at another Argos.[9] Homer writes: "And those who held Mycenae, well built fortress, and wealthy Corinth and well-built Cleonae, and dwelt in Orneia and lovely Araethree and Sicyon, wherein Adrastus was king…"[10]

The Goddess Demeter was called Melaina, the "Black One," and she was

[1] Ibid. VIII.2.1-32.
[2] Xenophon, *Anabasis* IV.8.2.
[3] Vitruvius, *On Architecture* VII.7.2.
[4] Strabo, *Geographica* XII.2.10.
[5] Dioscorides, *De Materia Medica* V.111; Pliny, *Natural History* XXXV.34-35.
[6] Procopius, *History of the Wars* VIII.2.4.
[7] Xenophon, *Anabasis* IV.8.20-21.
[8] Aristotle, *On Marvelous Things Heard* 831^{b}20.
[9] Strabo, *Geographica* VIII.6.19.
[10] Homer, *Iliad* II.569-571.

portrayed with a horse's head in wooden cult effigy.[1] According to legend, Demeter was said to have borne to Poseidon a steed named Arion, and Heracles gave Arion to Adrastus, son of Cretheus.[2] Homer writes in the *Iliad*: "divine Arion, the swift horse of Adrastus, who was of the race of the gods."[3] In his work *The Fall of Troy*, Quintus Smyrnaeus writes: "Agenor smote Molus the princely,—with king Sthenelus he came from Argos," adding, "And now the flashing-footed Argive steed by Sthenelus bestridden…Noble was he, for in his veins the blood of swift Arion ran…Him did the blessed Adrastus to give: And from him sprang the steed of Sthenelus, which Tydeus' son had given unto his friend in hallowed Troyland."[4] In the *Odyssey*, Homer writes of those who "perished in the broad land of Troy, far from horsepasturing Argos."[5]

Strabo says that Sarpedon brought the Leleges to Lycia (Asia Minor) from Crete.[6] In the *Iliad*, Homer says the god-like Sarpedon was the son of Zeus and Laodameia, the daughter of Bellerophon, and the poet considered them all to be native to Lycia.[7] According to Strabo, Homer identifies the Lycians as the Trojans when the poet writes, "'Trojans and Lycians and Dardanians,'—meaning that the Lycians and Dardanians were Trojans," adding, "the Leleges and Cilicians were so closely related to the Trojans…"[8] Strabo writes, "Now the poet (Homer) calls 'Trojans' the peoples, one and all, who fought on the Trojan side, just as he called their opponents both 'Danaans' and 'Achaeans.'"[9] The Trojans were a Greek nation that spoke Thracian and emigrated to the Troad (Northwest Asia Minor) from the Peloponnesus.[10]

According to Strabo, Sarpedon called the people he encountered in Lycia Termilae, who were formerly named Milyae, and even earlier the Solymi.[11] In the *Iliad*, Homer records that during the Trojan War the Lycian Bellerophon "fought with the glorious Solymi," and his son Isander was killed while

---

1 Pausanias, *Description of Greece* VII.25.8; VIII.5.8.
2 Ibid. VIII.25.9
3 Homer, *Iliad* XXIII.346.
4 Quintus Smyrnaeus, *The Fall of Troy* IV.563-573; VI.624-625.
5 Homer, *Odyssey* V.92-95.
6 Strabo, *Geographica* XII.8.5.
7 Homer, *Iliad* VI.99.
8 Strabo, *Geographica* XII.8.6.
9 Ibid. XIV.3.10.
10 Dionysius of Halicarnassus, *Roman Antiquities* I.61; Euripides, *Rhesus* 294.
11 Strabo, *Geographica* XII.8.5.

fighting the Solymi.[1] Strabo writes, "Now the poet (Homer) makes the Solymi different from the Lycians, for when Bellerophon was sent by the king of the Lycians to the second struggle, 'he fought against the glorious Solymi,'" adding, "by 'Solymi' the poet means the people who are now called Milyae..."[2]

Quoting Aristotle's pupil Choerilus, Josephus records that among the men marching with the Persian Xerxes against Greece (480 B.C.) were the Solymi: "Strangely upon their lips the tongue of Phoenicia sounded. In the Solomian hills by the broad lake their habitation, shorn in a circle, unkempt was the hair on their heads, and above them proudly wore their hides of horseheads dried in hearth-smoke."[3] The Jewish Josephus identifies the region as Lake Bitumen in the Solomian hills of Phoenicia (Palestine), making it "obvious" that Cheorilus was "referring to us," although Mosaic Law prohibited hair shorn in a circle.[4] According to the Latin historian Tacitus, there were some who believed that the Jewish people were of illustrious descent, and were, in fact, the Solymi who founded the city and gave it the name Hierosolyma (Jerusalem).[5] According the Plutarch, Cleombrotus told him that the Solymi, "who live next to the Lycians, paid especial honor to Cronos."[6]

In the *Odyssey*, Homer speaks of the Ethiopians from the mountains of the Solymi, who were called "White Ethiopians," and "White Syrians."[7] Herodotus states that the Ethiopians of Asia had straight hair, and "wore on their heads the skin of horses' foreheads, stripped from the head with ears and mane."[8] Strabo writes of the "White Syrians, whom we call Cappadocians," and says that these people bred horses and paid part of their tribute to Persia in horses.[9] Strabo quotes Herodotus who says that the Paphlagonians are bounded on the east by the Halys River, which "flows between the Syrians and the Paphlagonians and empties into the Euxine Sea (Black Sea)," and states, "by 'Syrians,' however, he means the Cappadocians, and in fact they

[1] Homer, *Iliad* VI.184-204.
[2] Strabo, *Geographica* XIV.3.10.
[3] Josephus, *Against Apion* I.173-174.
[4] *Leviticus* 19.27.
[5] Tacitus, *Histories* V.2.
[6] Plutarch, *De Defectu Oraculorum* 21.
[7] Homer, *Odyssey* V.295; Pliny, *Natural History* V.127.
[8] Herodotus, *Histories* VIII.70.
[9] Strabo, *Geographica* XI.13.8.

are still today called 'White Syrians,' which those outside the Taurus are called Syrians."[1] He adds, "those outside have a tanned complexion, while those this side do not, and for that reason received the appellation 'white.' And Pindar says the Amazons 'swayed a "Syrian" army that reached afar with their spears,' this clearly indicating that their abode was in Themiscrya. Themiscyra is in the territory of the Amiseni, and this territory belongs to the White Syrians."[2]

According to Herodotus, marching in Xerxes army were "the Cabalees, who are Meiones, who are called Casonii," and also named Milyae, and they "carried Lycian bows and wore caps of skins on their heads."[3] Herodotus writes, "What is now possessed by the Lycians was of old Milyan and the Milyans were then called Solymi."[4] According to Strabo, "the Solymi occupied the loftiest peaks of the Taurus range, I mean the peaks about Lycia as far as Pisidia."[5] Strabo says that Homer's words, "on his way from the Ethiopians he espied Odysseus from the mountains of the Solymi," is equivalent to saying "from the region of the south," and Strabo insists that Homer "does not mean the Solymi of Pisidia."[6]

According to Strabo, Termessus was a Pisidian city of Phrygia that was held to be Lycian, adding, "the Cabaleis are said to be the Solymi, at any rate, the hill that lies above the fortress of the Termessians is called Solymus, and the Termessians themselves are called Solymi."[7] The Lycian Leleges intermarried with the Solymi.[8] Consequently, Eratosthenes listed the Solymi and Leleges among the races extinct.[9]

The Leleges were Carians, and they believed themselves to be brethren of the Thracian Mysians and Lydians.[10] Strabo says that the Solymi were identical to the Mysians, the Lydians and the Meiones.[11] Lydia was originally called Maeonia, and this was bordered by Phrygia to the east, Mysia to the

[1] Ibid., XI.14.9.
[2] Ibid., XII.3.5-9.
[3] Herodotus, *Histories* VII.77.
[4] Ibid. I.173.
[5] Strabo, *Geographica* I.2.10.
[6] Ibid. I.2.28.
[7] Ibid. XIII.4.6; XIV.3.9.
[8] Ibid. XII.7.3.
[9] Pliny, *Natural History* V.127.
[10] Herodotus, *Histories* I.177; Strabo, *Geographica* VII.7.2.
[11] Strabo, *Geographica* XII.3.20; XIII.3.2.

north, and Caria to the south.[1] Strabo writes, "The parts situated next to this region towards the south as far as the Taurus are so interwoven with one another that the Phrygian and the Carian and the Lydian parts, as those of the Mysians, since they merge into one another, are hard to distinguish."[2]

The Lydians and the Mysians occupied the territory of Phrygia "Catececaumene", and the Catececaumene country was called Mysia and Meionia.[3] According to Strabo, some place in this region the mythical story of Typhon, and Xanthus says that Arimus was the king of this region.[4] The Greek poet Pindar writes, "Typhon with his hundred heads, who was nurtured of old by the famed Cilician cave…"[5] Curtius writes, "In that region lapse of time had destroyed many memorials made famous in song: the sites of the cities Lyrnesus and Thebes were pointed out, the cave of Typhon too, and the Corycian grove, where saffron grows, and other places of which only the fame has endured."[6] Quoting Homer, Strabo writes, "in the land of the Arimi where men say is the couch of Typhon," adding that Demetrius "thinks that those writers are most plausible who place the Arimi in the Catececaumene country in Mysia."[7] Strabo says that the Arimi are now called Arimaeans, and he quotes Callisthenes who places the Arimi in the mountains called Arimi, "near Mount Calycadnus and the promontory of Sarpedon near the Corycian cave itself."[8]

Strabo says that Homer mentions the "Arami," and according to Poseidonius this should be interpreted as meaning Syria itself, because the people of Syria are called Aramaeans, and the Greeks changed this to Arimaeans or Arimi. According to Strabo, the Arimi were also called Arimeans and Arammeans, and they inhabited Mesopotamia (Assyria) along with the Arabs and Armenians, declaring, "the Assyrians, the Arians, and the Arammeans display a certain likeness both to those just mentioned and to one another."[9]

[1] Pliny, *Natural History* V.110.
[2] Strabo, *Geographica* XIII.4.2.
[3] Ibid. XII.8.18.
[4] Ibid. XIII.4.11.
[5] Pindar, *Pythian Odes* I.30.
[6] Quintus Curtius, *History of Alexander* III.4.10.
[7] Strabo, *Geographica* XIII.4.6.
[8] Ibid.
[9] Strabo, *Geographica* I.2.34; Homer, *Iliad* II.783.

Strabo again quotes Homer who says that the Meionians were "born at the foot of Tmolus," and this mountain is in the territory of the Lydians.[1] Strabo writes "Hypaepa is a city which one comes to on the descent from Mount Tmolus to the Caystar Plain," and he quotes Callinus, who says that Meionia "was called Asia, and accordingly Homer likewise says 'on the Asian mead about the streams of the Caystar.'"[2] Silius writes, "And when the sun rose and the hoofs of Phaeton's horses dispelled the dews, all Mount Massicus was green with vine-bearing fields, and marveled at the leafage and the bunches shining in the sunlight. The fame of the mountain grew, and from that day fertile Tmolus and the nectar of Ariusia and the strong wine of Methymna have all yielded precedence to the vats of Falernus."[3]

The Argo was a Thessalian ship,[4] and according to Strabo, the city of Pherae on the Pelasgian plain was only ninety stadia from Iolcus, where Pelias dispatched Jason and the Argo.[5] Strabo writes:

> As for the Pelasgi, almost all agree in the first place, that some ancient tribe of that name spread through the whole of Greece, and particularly among the Aeolians of Thessaly. Again, Ephorus says that he is of the opinion that, since they were originally Arcadians, they chose a military life, and that, in converting many peoples to the same mode of life, they imparted their name to all, and thus acquired great glory...For example, they proved to have been the colonizers of Crete...and Thessaly is called "the Pelasgian Argos"... and Homer has called the Pelasgi the people that were neighbors to those Cilicians who live in the Troad...[Aeschylus] says that the race of the Pelasgi originated in that Argos which is round Mycenae...[some] sailed to Italy with Tyrrhenus the son of Atys.[6]

Near Lake Copais in Boeotia (Greece) was the settlement Orchomenus, and according to Strabo, "the poet (Homer) gives the catalog of the Orchomenians, whom he separates from the Boeotian tribe. He calls Orchomenus 'Minyeian,' after the Minyae. They say that some of the Minyae

[1] Strabo, *Geographica* XIII.4.12.
[2] Ibid. XIII.4.7.
[3] Silius, *Punica* VI.205-211.
[4] Plautus, *Amphitryon* 1043-1046; Propertius I.20.17-20.
[5] Strabo, *Geographica* XI.5.15.
[6] Ibid. V.2.4.

emigrated from here to Iolcus, and that from this fact the Argonauts were called Minyae."[1] Herodotus declares that the Thessalians named Minyeian Orchomenians, or Minyae, were the "descendants of the heroes who sailed the Argo, and put in at Lemnos (Greek island off Euboea) and there begotten their race."[2] According to Strabo, beyond Armenia near the Medes (Persians) and Caspian Sea there settled a tribe of Thracians, "and therefore from all this it is supposed that both the Medes and Armenians are in a way kinsmen to the Thessalians, and the descendants of Jason and Medea."[3]

Strabo places other tribes of Thracians and Thracian Euboeans north of the Black Sea:

> The Amazons also are said to live in the mountains above Albania...the Gelae and Legae, Scythian people, live between the Amazons and the Albanians...[others say] that the Amazons live on the borders of the Gargarians, in the northerly foothills of those parts of the Caucasias Mountains...They have two special months in the spring in which they go up into the neighboring mountains which separates them from the Gargarians. The Gargarians also, in accordance with an ancient custom, go up thither to offer sacrifice with the Amazons and also to have intercourse with them...the females that are born are retained by the Amazons themselves, but the males are taken to the Gargarians to be brought up...It is said that the Gargarians went up from Themiscryra into the region with the Amazons, and revolted from them in company with some Thracians and Euboeans who had wandered thus far carried on war against them, and that they later ended the war against them and made a compact on the conditions mentioned above...[4]

Settling in northern Europe, the Euboeans, Graeco-Thracian descendants of the Argonauts, became known as the Argippeans, a sacred people who utilized a plant called "Pontic," a name for Asia Minor, indicating their Phrygian origins. According to Herodotus, there was limited information available about the tribes of northern Europe, even though they were in contact with the Scythian (German) tribes and the Greeks of Asia Minor: "Now as

[1] Ibid. XI.2.40.
[2] Herodotus, *Histories* IV.145.
[3] Strabo, *Geographica* XI.14.14.
[4] Ibid. XI.5.1-2.

far as the land of these bald men (Argippeans) we have full knowledge of the country and the nation on hither side of them; for some of the Scythians make their way to them, from whom it is easy to get knowledge, and from some too of the Greeks from the Borysthenes port and other ports of Pontus; such Scythians as visit them do their business with seven interpreters and in seven languages."[1] He writes:

> [The Argippeans] speak a tongue of their own, and wear Scythian raiment, and their fare comes from trees. The tree wherefrom they live is called "Pontic"; it is about the size of a fig tree, and bears a fruit as big as a bean, with a stone in it, which they call "aschu"; they lick this up or mix it with milk for drinking, and the thickest of the lees of it they make cakes and eat them…These people are wronged by no man, for they are said to be sacred; nor have they any weapon of war. These are they who judge in the quarrels between their neighbors; moreover, whatever banished man has taken refuge with them is wronged by none.[2]

According to Pliny, near the Rhipaean Mountains dwell the Arimphaei ("One-eyed men), "a race not unlike the Hyperboreans. They dwell in forests and live on berries…their manners are mild. Consequently they are reported to be deemed a sacred tribe and are left unmolested even by the savage tribes among their neighbors, this immunity not confined to themselves but extended also to people who have fled to them for refuge."[3] Ammianus writes, "where the Riphaean mountains sink to the plain, dwell the Arimphaei, just men known for their gentleness."[4]

According to Herodotus, the poets of the eighth and ninth century B.C. were aware of the Hyperboreans: "Concerning the Hyperborean people, neither the Scythians nor any other dwellers in these lands tell us anything, except perchance the Issedones. And, as I think, even they tell nothing; for were it not so, then the Scythians too would have told, even as they tell of the one-eyed men. But Hesiod speaks of the Hyperboreans, and Homer too in his poem *The Heroes' Son*, if that be truly the work of Homer."[5] Citing

[1] Herodotus, *Histories* IV.24.
[2] Ibid. IV.28.
[3] Pliny, *Natural History* IV.23.
[4] Ammianus Marcellinus, *The Chronicles of Events* XXII.8.38.
[5] Herodotus, *Histories* IV.32.

Hellanicus (5th century B.C.) as his authority, Clement places the vegetarian Hyperboreans in the land of the one-eyed men, "Hellanicus says that the Hyperboreans live beyond the Rhipaean Mountains. They learn justice, and eat no meat but live from berries."[1] According to the Latin historian Aelian, the Hyperboreans were said to be the happiest of all men.[2] Herodotus says that the Hyperboreans were a peaceful people: "Except for the Hyperboreans, all these nations (of the region) ever make war upon their neighbours."[3] According to Diodorus, the Hyperboreans had their own language, "and are most friendly disposed towards the Greeks, and especially towards the Athenians and the Delians, who inherited this good-will from most ancient times."[4] In his *Pythian Odes*, Pindar reveals that the Hyperboreans were a chosen people who honored Apollo:

> Neither by ships nor by land canst thou find the wondrous road to the trysting-place of the Hyperboreans. Yet among them, in olden days, Perseus, the leader of the people, shared the banquet on entering their homes and finding them sacrificing famous hecatombs of asses in honor of the god. In the banquets and praises of that people Apollo chiefly rejoiceth, and he laugheth as he looketh on the brute beast in their rampant lewdness. Yet, such are their ways that the Muse is not banished, but, on every side, the dances of maidens and the sounds of the lyre and the notes of the flute are ever circling; and with their hair crowned with golden bay-leaves, they hold glad revelry; and neither sickness nor baneful eld mingleth among that chosen people; but, aloof from toil and conflict, they dwell far from the wrath of Nemesis. To that host of happy men, went of old the son of Danaë…[5]

Apollo was surnamed Egyptian, and was identified by the Egyptians with the sun and Horus, and he was also considered the inventor of the seven-string cithara and aulos.[6] According to Herodotus, "it is said that the sacred objects sent by the Hyperboreans were in ancient times conducted to Delos to

[1] Clement of Alexandria, *Stromata* I.72.
[2] Aelian, *Various History* III.18.
[3] Herodotus, *Histories* IV.13.
[4] Diodorus Siculus, *The Library of History* II.47.4.
[5] Pindar, *Pythian Odes* X.25-45.
[6] Pausanias, *Description of Greece* II.27.6; Macrobius, *Saturnalia* I.21; Plutarch, *Moralia* 1132-42.

the music of auloi, of pipes of Pan, and of the cithara."[1] Diodorus associates Apollo with the Hyperboreans:

> [Apollo] broke the strings of the lyre and destroyed the harmony of sound which he had discovered. This harmony of strings, however, was rediscovered, when the Muses added latter the middle string, Linus the string struck with the forefinger, and Orpheus and Thamyrus the lowest string and the one next to it. And Apollo, they say, laid away both the lyre and the pipes as a votive offering in the cave of Dionysus, and becoming enamoured of Cybelê joined in her wanderings as far as the land of the Hyperboreans.[2]

According to Diodorus, the Hyperboreans inhabited cities on an island in northern Europe, and they were priests of Apollo who continually played the cithara and sang hymns of praise to the god:

> Hecataeus and certain others say that in the region beyond the land of the Celts [i.e. Gaul] there lies in the ocean an island no smaller than Sicily. This island [Britain], the account continues, is situated in the north and is inhabited by the Hyperboreans...Apollo is honored among them above all other gods, and the inhabitants are looked upon as priests of Apollo, after a manner, since daily they praise this god continuously in song and honour him exceedingly. And there is also on the island both a magnificent sacred precinct of Apollo and a notable temple which is adorned with many votive offerings and is spherical in shape [Stonehenge]. Furthermore, a city is there which is sacred to this god, and the majority of its inhabitants are players of the cithara; and these continually play on this instrument in the temple and sing hymns of praise to the god, glorifying his deeds. The Hyperboreans also have a language which is peculiar to them, we are informed, and are most friendly disposed towards the Greeks, who have inherited this good-will from most ancient times.[3]

According to Pliny, the Hyperboreans were a deeply religious people with ancient connections to Greece, and they regularly sent the "first fruits" of

[1] Herodotus, *Histories* II.83.
[2] Diodorus Siculus, *The Library of History* III.59.5-6.
[3] Ibid. II.47.1-4.

their harvest to Delos as offerings to Apollo:

> Along the coast, as far as the river Don, are the Maeotae from whom the sea receives its name, and last of all in the rear of the Maeotae are the Arimaspi. Then come the Ripaean Mountains and the region called Pterophorus...Behind these mountains and beyond the north wind there dwells (if we can believe it) a happy race of people called Hyperboreans, who live to extreme old age and are famous for legendary marvels...The homes of the natives are woods and groves; they worship the gods severally and in congregations, all discord and all sorrow is unknown. Death comes to them only when, owing to a satiety of life, after holding a banquet and anointing their old age with luxury, they leap from a certain rock into the sea; this mode of burial is the most blissful. Some authorities have placed these people not in Europe but on the nearest part of the coasts of Asia, because there is a race there with similar customs and a similar location, named the Attaci...Nor is it possible to doubt about this race, as so many authorities state that they regularly send the first fruits of their harvests to Delos as offerings to Apollo, whom they specially worship. These offerings used to be brought by virgins, who for many years were held in veneration and hospitably entertained by the nations on the route, until because of a violation of good faith they instituted the custom of depositing their offerings at the nearest frontiers of the neighbouring people, and these of passing them on to their neighbours, and so till they finally reached Delos. Later this practice itself also passed out of use.[1]

Citing the inhabitants of Delos as his authority, Herodotus says that the Hyperborean offerings were wrapped in wheat straw, and passed through the Scythian nations until they reached the Adriatic Sea, and were then passed to the Greek people of Dodona down to the Melian Gulf, then across to Euboea, and then from city to city the inhabitants carried these offerings on to Delos. He writes, "I can say of my own knowledge that there is a custom like these offerings, namely, when the Thracian and Paeonian women sacrifice to the Royal Artemis, they have wheat-straw with them while they sacrifice."[2]

[1] Pliny, *Natural History* IV.88-91.
[2] Herodotus, *Histories* IV.33.

According to Pausanias, the Athenians were the last people to receive the Hyperboreans' offerings to Apollo, the mysterious "first fruits" of the harvest hidden in wheat straw, and they took these "fruits" to Delos: "At Prasiae (in Greece) is a temple of Apollo. Hither they say are sent the first-fruits of the Hyperboreans, and the Hyperboreans are said to hand them over to the Arimaspi, the Arimaspi to the Issedones, from these the Scythians bring them to Sinope, thence they are carried by the Greeks to Prasiae, and the Athenians take them to Delos. They first-fruits are hidden in wheat straw, and they are known of none."[1] The Greek scholar Callimachus (3rd century B.C.) tells a similar story:

> ...[lacuna]...yearly, along with the tribute of the tenth... [lacuna]...they [send] the divine planks...[lacuna]...the sons of the Hyperboreans send from the Rhipaean Mountains where the rich sacrifice of donkeys pleases Phoebus [Apollo] particularly. Of the Greeks the Pelasgian Ellopians [Dodonaeans] first accept these conveyed by the Arimaspi...from there the servants of Zeus Naios with unwashed feet send them to the cities and the mountains of the land of Malis...[lacuna]...[2]

Buried in the precinct of Artemis at Delos were Hyperoche and Laodice, two virgin maidens of the Hyperboreans who were sent with the offerings to Apollo but never returned home.[3] According to Herodotus, the children of Delos honored these maidens, apparently the victims of a great tragedy:

> But on the first journey the Hyperboreans sent two maidens bearing the offerings, to whom the Delians give the names Hyperoche and Laodice, sending with them for safe conduct five men of the people as escort, those who are now called Perpherees and greatly honoured at Delos. But when the Hyperboreans found that these whom they sent never returned, they were very ill content that it should ever be their fate not to receive their messengers back; wherefore they carry the offerings, wrapt in wheat-straw, to their borders, and charge their neighbours to send them on from their own country to the next...The Delian girls and boys cut their hair

[1] Pausanias, *Description of Greece* I.31.2.
[2] Callimachus, *Aetia* 186.
[3] Eusebius, *Praeparatio Evangelica* II.6.

> in honour of the Hyperborean maidens, who died at Delos, the girls before their marriage cut off a tress and lay it on the tomb wound about a spindle; this tomb is at the foot of an olive-tree, on the left hand of the entrance of the temple of Artemis; the Delian boys twine some of their hair around a green stalk, and they likewise lay it on the tomb.[1]

According to Herodotus, before the deaths of Hyperoche and Laodice, the two Hyperborean virgin maidens, Arge and Opis, came to Delos to bring tribute to the Hyperborean Eileithyia for something to ease childbirth, and the Greeks came to celebrate these virgin maidens:

> These same Delians relate that two virgins, Arge and Opis, came from the Hyperboreans by way of the peoples aforesaid to Delos, yet earlier than the coming of Hyperoche and Laodice; these latter came to bring to Ilithyia the tribute whereto they had agreed for ease of child-bearing; but Arge and Opis, they say, came with the gods themselves, and received honour of their own from the Delians. For the women collected gifts from them, calling upon their names in the hymn made for them by Olen, a man of Lycia; it was from Delos that the islanders and Ionians learnt to sing hymns to Opis and Arge…[2]

Pausanias says that at Athens there is "a temple of Eileithyia, who they say came from the Hyperboreans to Delos and helped Leto in her labour; and from Delos the name spread to other peoples. The Delians sacrifice to Eileithyia and sing a hymn of Olen."[3] According to Pausanias, Olen composed a hymn to Eileithyia, comparing her to the Great Mother Goddess and calling her the mother of the Goddess Love:

> Most men consider Love to be the youngest of the gods and the child of Aphrodite. But Olen the Lycian, who composed the oldest Greek hymns, says in a hymn to Eileithyia that she was the mother of Love. Later than Olen, both Panphos and Orpheus wrote hexameter verse, and composed poems on Love, in order that they might be among those sung by the Lycomidae to accompany the ritual. I

[1] Herodotus, *Histories* IV.33-34.
[2] Herodotus, *Histories* IV.35.
[3] Pausanias, *Description of Greece* I.18.5.

> read them after conversation with a Torchbearer. Of these things I will make no further mention...The first to remove the image of Love, it is said, was Gaius the Roman Emperor; Claudius, they say, sent it back to Thespiae, but Nero carried it away a second time. At Rome the image perished by fire. Of the pair who sinned against the god, Gaius was killed by a private soldier, just as he was giving the password; he had made the soldier very angry by always giving the same password with a covert sneer. The other, Nero, in addition to the violence to his mother, committed accursed and hateful crimes against his wedded wives.[1]

According to Pausanias, at one of the temples of Apollo in Greece, "Oracular responses are still given here, and the oracle acts in the following way. There is a woman who prophesies, being debarred from intercourse with a man. Every month a lamb is sacrificed at night, and the woman, after tasting the blood, becomes inspired by the God."[2] He also declares that it was Apollo who inspired prophetic oracles: "those whom they say Apollo inspired of old none of the seers uttered oracles, but they were good at explaining dreams and interpreting the flights of birds and the entrails of victims."[3] At Delphi in Greece the foremost priestly family was the Thracidae, clearly the descendants of the Thracians, and Plutarch states that the oracle at Delphi was "the most ancient in time and the most famous in repute...they used to employ two prophetic priestesses who were sent down in turn; and a third was appointed to be held in reserve."[4] Plato records that in the temple of Apollo at Delphi there were the inscriptions "Know thyself" and "Avoid extremes."[5] Pausanias says that the Hyperboreans established the oracle of Apollo at Delphi, and that the Hyperborean Olen was the first prophet of the temple:

> The most prevalent view, however, is that Phemonoë was the first prophetess of the god [Apollo], and first sang in hexameter verse. Boeo, a native woman who composed a hymn for the Delphians, said that the oracle was established for the god by comers from the

[1] Ibid. IX.27.2-4.
[2] Ibid. II.24.1.
[3] Ibid. I.34.4.
[4] Plutarch, *Moralia* 414.
[5] Plato, *Protagorus* 343B; etc.

> Hyperboreans, Olen and others, and that he was the first to prophesy and the first to chant the hexameter oracles. The verses of Boeo are:—"Here in truth a mindful oracle was built by the sons of the Hyperboreans, Pagasus and divine Agyieus." After enumerating others also of the Hyperboreans, at the end of the hymn she names Olen:—"And Olen, who became the first prophet of Phoebus, and first fashioned a song of ancient verses." Tradition, however, reports no other man as prophet, but makes mention of prophetesses only. They say that the most ancient temple of Apollo was made of laurel... The Delphians say that the second temple was made by bees from bees-wax and feathers, and that it was sent to the Hyperboreans by Apollo.[1]

The legendary travels of the Hyperborean Abaris was recorded by several ancient writers, and some have it that the wandering Druid was taught by Pythagoras when the priest of Apollo arrived in Italy.[2] Aristotle says that Pythagoras was called Hyperborean Apollo in Croton.[3] In his *De Vita Pythagorica*, Iamblichus writes: "For Abaris came from the Hyperboreans, a priest of Apollo there, an old man and most wise in sacred matters...Abaris agreed to remain with him," adding, "[Pythagoras] taught him his doctrine of nature and theology in abridged form. Instead of divination by the entrails of sacrificed animals, he taught him fore-knowledge through numbers, believing this to be purer, more divine, and more suitable to the heavenly numbers of the gods; and he also taught other studies suitable to Abaris' interests."[4]

Strabo writes, "[Poseidonius] says that there is an island near Britain on which sacrifices are performed like those sacrifices in Samothrace that have to do with Demeter and Corê," adding, "Many writers have identified the gods that are worshipped in Samothrace with the Cabeiri, though they cannot say who the Cabeiri themselves are..."[5] Pausanias speaks of the "Celts who live farthest off on the borders of the land which is uninhabited because of the cold; these people, the Cabares..."[6] Again citing Poseidonius as his authority,

[1] Pausanias, *Description of Greece* X.5.6-9.
[2] Herodotus, *Histories* IV.36; etc.
[3] Aelian, *Various History* II.26.
[4] Iamblichus, *De Vita Pythagorica* 90-93.
[5] Strabo, *Geographica* IV.4.6; *Fragment* VII.50.
[6] Pausanias, *Description of Greece* I.35.5.

Strabo records the existence of an Atlantic island inhabited by women who were possessed by Dionysus, also performing mystic initiations, and once a year they would sacrifice a victim with the Thraco-Phrygian cry of "Ev-ah":

> In the [Atlantic] ocean, he [Poseidonius] says, there is a small island, not very far out to sea, situated off the outlet of the Liger River; and the island is inhabited by the women of the Samnitae, and they are possessed by Dionysus and make this god propitious by appeasing him with mystic initiations as well as other sacred performances; and no man sets foot on the island, although the women themselves, sailing from it, have intercourse with the men and then return again. And, he says, it is a custom of theirs once a year to unroof the temple and roof it again on the same day before sunset, each woman bringing her load to add to the roof; but the woman whose load falls out of her arms is rent to pieces by the rest, and they carry the pieces round the temple with the cry of "Ev-ah," and he says, that some one jostles the woman who is to suffer this fate.[1]

Clearly revealing these people to be identical with the Phrygian Dactyli who were credited with the Olympic games, Pausanias says that Heracles brought the Olympic games into Greece from the land of the Hyperboreans.[2] He writes, "Many are the sights to be seen in Greece, and many are the wonders to be heard; but on nothing does Heaven bestow more care than on the Eleusinian rites and the Olympic games."[3]

The similarities between the Thraco-Phrygian and British people cannot be dismissed. Like the Thracians the British men had a reputation for being fierce, formidable, brave, and extremely good fighters.[4] And like the Thracians the British also used chariots in war.[5] Like the Thracians the British tribes practiced tattooing.[6] Like the Thracians, the British tribes wore their hair

[1] Strabo, *Geographica* IV.4.6.

[2] Pausanias, *Description of Greece* V.7.7.

[3] Ibid. V.10.1.

[4] Catullus, XII.II; Herodian, *History of the Roman Empire* II.15.1; III.7.2; Claudian, *Gothic War* 568.

[5] Strabo, *Geographica* IV.5.2; Diodorus Siculus, *The Library of History* V.21.5; A. Fol and I. Marazov, *Thrace & the Thracians*, New York, St. Martin's Press, 1977, pp. 50, 106-107.

[6] Herodian, *History of the Roman Empire* III.14.6-7; Oppian, *Cynegetica* I.470.

long, and like the Thracians they were polygamous in their relationships.[1] According to Diodorus, the British "are simple and far removed from the shrewdness and vice with characterize the men of our day. Their way of life is modest, since they are well clear of the luxury which is begotten of wealth."[2] But most revealing is the religion of the British, and their priests, the Druids, or "Holy Ones," who were indigenous to the British, and were also found among the Celts in Gaul (the home of the Bretons), and in Ireland.[3]

The British (Cornish, Welsh, Breton) Druids were judicial judges, philosophers in religious affairs, and parishioners of astrology, and they were exempt from taxes, military service, and all other liabilities, allowing them to perform human sacrifices, customarily the judicial execution of a convicted criminal.[4] The Druids taught that the soul is immortal, and indestructible like the entire universe.[5] According to Ammianus, the Phocians emigrated from Greece and settled in Massilia (Marseilles, in Gaul), where they became civilized through the Pythagorean (likely from Abaris) liberal arts, initiated by the Druids, who were "loftier than the rest in intelligence, and bound together in fraternal organizations, as the authority of Pythagoras determined, were elevated by the investigation of obscure and profound subjects, and scorning all things human, pronounced the soul immortal."[6]

The Druids did not think it was proper to commit their teachings to writing, and therefore their tradition was transmitted orally, although these priests were literate, and when they did write they used Greek letters.[7] According to Diogenes, the "Druids we are told that they uttered their philosophy in riddles, bidding men to reverence the gods, abstain from wrongdoing, and to practice courage."[8] In his work *The Professors at Bordeaux*, Ausonius of Bordeaux says that the Druids worshipped Belanus (Belus), who he identifies as Apollo, and speaking of Patera the Elder, he writes: "Your father and your brother were named after Phoebus (Apollo), and your son after Delphi…Nor

[1] Julius Caesar, *Gallic War* V.14.
[2] Diodorus Siculus, *The Library of History* V.21.6.
[3] Diogenes Laertius, *Lives of Eminent Philosophers* I.1.
[4] Diodorus Siculus, *The Library of History* V.32.2; Tacitus, *The Annals* XIV.30; Julius Caesar, *Gallic Wars* VI.13-14.
[5] Strabo, *Geographica* IV.4.4; Lucan, *Pharsalia* I.449-463.
[6] Ammianus Marcellinus, *The Chronicles of Events* XV.9.7-8.
[7] Julius Caesar, *Gallic War* VI.14.
[8] Diogenes Laertius, *proem* 6.

must I leave unmentioned the old man Phoebicius, who though the keeper of Belanus' temple, got no profits thereby. Yet he sprung, as rumour goes, from the stock of the Druids of Armorica (Brittany, in Gaul)…"[1]

In his *Hymn to Apollo*, Callimachus says that the youthful Apollo's "locks distil fragrant oils upon the ground; not oil of fat do the locks of Apollo distil, but the very Healing of All (*Panaceas*). And in whatever city those dews fall upon the ground, in that city all things are free from harm."[2] According to Theophrastus, the panacea is "like the moly mentioned in Homer. It is used against spells and magic arts, but that it is not as Homer says, difficult to dig up."[3] The panacea was also named asclepion and heracleon, and was considered a universal remedy for all illness, and this was an odiferous plant that grew in the mountains, and the kind imported from Macedonia was called bucolion (pastorial) because the herdsmen collect it as it exuded of its own accord.[4] The code of ethics that physicians are charged with upholding when they begin to practice medicine is contained in the Hippocratic *Oath*, which begins: "I swear by Apollo the physician, by Asculepius, by Hygeia, by Panacea and all the gods and goddesses…"[5]

Pliny calls the Druids "magicians" (*magos*), and says that their name is derived from the Greek word for "oak," and these priests believed that anything growing on this tree was sent from heaven, and was a sign that the tree was chosen by God himself.[6] According to Pliny, the oak produces edible fungi and hog mushrooms that grow around its roots, and it also produces mistletoe, held to be sacred, especially when two white bulls are sacrificed, "and the magicians (Druids) perform no rites without using the foliage of those trees."[7] Athanaeus writes, "Ion, in *Phoenix* or *Carneus*, called mistletoe the 'sweat of the oak.'"[8] According to Plutarch, honey from the oak was used to prepare a certain drink for men.[9] Medea was associated with the magical root found on the oak, which produced a bellowing (*mykethmos*, Greek *mykes*,

[1] Ausonius of Bordeaux, *The Professors at Bordeaux* IV.6; X.22.
[2] Callimachus, *Hymn to Apollo* 38-40.
[3] Theophrastus, *History of Plants* IX.15.5.
[4] Pliny, *Natural History* XIX.165; XX.31; XXV.30-33; Theophrastus, *History of Plants* IX.15.5.
[5] Hippocrates, *Oath* 1-3.
[6] Pliny, *Natural History* XVI.249-251.
[7] Ibid. XVI.31.
[8] Athenaeus, *Deipnosophistae* X.451.
[9] Plutarch, *Coriolanus* 3.

"mushroom") sound that shook the earth when she cut it to anoint Jason in the Garden of Colchis.[1] According to Antiphanes, there was a mushroom found on the oak that could induce clairvoyance.[2] Mistletoe was a drug that was used medically for epilepsy, sores, wounds, and swellings among other ailments.[3]

According to Pliny, the oak produces the best honey.[4] In his *Metamorphoses*, the Latin poet Ovid writes: "Anon, the earth, untilled, brought forth her store of grain, and the fields, though unfallowed, grew white with the heavy bearded wheat. Streams of milk and streams of nectar flowed, and yellow honey was distilled from the verdant oak."[5] The Latin poet Virgil writes: "slowly shall the plain yellow with the waving corn, on wild brambles shall hang the purple grape, and the stubborn oak shall distil dewy honey."[6] In his *Alexipharmaca*, Nicander writes, "Bees were born from the carcass of a calf that had fallen in the glades, and there in some hollow oak they first, maybe, united to build their nest, and then, bethinking themselves of work, wrought round it in Demeter's honour their many-celled combs…"[7]

In his work entitled *On the Cave of the Nymphs*, Porphyry says some equate honey with the ambrosia and call it the "food of the gods," and he also says that the priestesses of Demeter were called "mystic bees," and the ambrosia was mixed in amphoras by these "bees."[8] This is reminiscent of Pindar who writes in his *Pythian Odes*: "they shall drip nectar and ambrosia on his lips and shall make him immortal."[9] Pliny writes, "For after the rising of each star, but particularly the principal stars…if rain does not follow but the dew is warmed by the rays of the sun, not honey but drugs are produced, heavenly gifts…its sweetness and potency for recalling mortal's ills from death is equal to that of the nectar of the gods."[10] Theophrastus contends that the oak produces fungi and mistletoe that also grows on other trees, but "the oak…produces more things than any other tree; and all the more so, if, as

[1] Apollonius Rhodius, *Argonautica* III.3-20; III.858-880.
[2] Antiphanes, *Fragment* 227.
[3] Dioscorides, *De Materia Medica* III.103; Pliny, *Natural History* XVI.249-251.
[4] Pliny, *Natural History* XI.32.
[5] Ovid, *Metamorphoses* I.108-112.
[6] Virgil, *Ecologues* IV.27-30.
[7] Nicander, *Alexipharmaca* 447-452.
[8] Porphyry, *On the Cave of the Nymphs* 15-18.
[9] Pindar, *Pythian Odes* IX.63.
[10] Pliny, *Natural History* XI.37.

Hesiod says, it produces honey and even bees; however, the truth appears to be that this honey-like juice comes from the air and settles on this more than on other trees."[1] Pliny writes, "it is an accepted fact that honeydew falling from the sky, as we have said, deposits itself on no other tree in preference to the hard oak…[this will] produce agaric, which is a white fungus with a strong odor, and which makes a powerful antidote; it grows on the tops of trees, and is phosphorescent at night; this is its distinguishing mark, by which it can be gathered in the dark."[2]

[1] Theophrastus, *History of Plants* IV.6.7.
[2] Pliny, *Natural History* XVI.31-32.

CHAPTER 2: INTRODUCTION TO THE SOMA ENIGMA

> Then spake Zarathustra: Reverence to Homa! good is Homa, well-created is Homa, rightly created, of a good nature, healing, well-shaped, well-performing, successful, golden-coloured, with hanging tendrils, as the best for eating and the most lasting provision for the soul.[1]

The *Ṛgveda* is a collection of religious hymns originally guarded as a secret text, composed by the Vedic priests of the Indo-Aryans who migrated into India around the beginning of the second millennium B.C. The *Ṛgveda* is the oldest written record of the Indo-Aryans, and is religiously and linguistically connected with the Persian *Avesta*, containing the sacred writings of Zarathustra and the Zoroastrian religion.[2] In tedious and complicated rituals, the ancient Vedic priests prepared and ingested an extraordinary plant with an effect they equated with immortality, called Soma in the *Ṛgveda*, and Haoma in the *Avesta*. Soma is venerated and deified in the *Ṛgveda* as the

[1] † The sources here are quoted verbatim, and therefore there are minor variations in punctuation and the spelling or italicizing of particular words. For the sake of accuracy, these variants are presented in the form in which they appear in the original source.

Hōma Yasht IX.16; in Martin Haug, *Essays on the Sacred Language, Writings, and Religion of the Parsis*, Popular Edition, London, Kegan Paul, Trench, Trübner & Co., 1883, pp. 179-180.

[2] Āthrava is the name for "priest" in the *Avesta*, and Atharvan means priest of fire and Soma in the Vedas. Ibid., p. 280.

"Ruler of the plants,"[1] the "Lord of the mind,"[2] and a "God, pressed out for Deities."[3] Repeating a verse from the *Ṛgveda*, the Vedic priests performing the Soma-sacrifice would recite the mantra: "We have drunk Soma, we have become immortal, we have entered into light, we have known the gods."[4] Asko Parpola writes:

> Indra is the most popular deity of the Ṛgveda, the god of war and thunder and a central element in Indra's cult was a drink originally called **Sauma*: Vedic *Soma*, and Avestan *Haoma*, the cultic drink which Zoroastrianism evidently adopted from the earlier Bronze Age religion of Central Asia and eastern Iran. Indra was undoubtedly associated with *Haoma* also in this religion against which Zarathustra rebelled — Indra is invoked by the Mitanni Aryans in 1380 B.C. — but he was dethroned and made a demon by Zarathustra…*Soma* is likely to be a Proto-Indo-Aryan invention.[5]

Several distinguished Vedic scholars of the nineteenth and early twentieth centuries were of the opinion that Soma was a fermented beverage; F. Max Müller,[6] Rájendralála Mitra,[7] Abel Bergaigne,[8] Adolf Kaegi,[9] Ralph T. H. Griffith,[10] Arthur Anthony Macdonell,[11] Arthur Berriedale Keith,[12] and E.

---

[1] *Ṛgveda* IX.114.2; in Ralph T. H. Griffith (trans.), *The Hymns of the Rgveda*, Second Edition, Volumes I-II, Benares, E. J. Lazarus, 1889-1892, vol. II, p. 382.

[2] *Ṛgveda* IX.28.1; in Ibid., vol. II, p. 289.

[3] *Ṛgveda* IX.3.9; in Ibid., vol. II, p. 271.

[4] *Ṛgveda* VIII.48.3; in Arthur Anthony Macdonell, *Vedic Mythology*, Strassburg, Karl J. Trübner, 1897, p. 109.

[5] Asko Parpola, "The problem of the Aryans and the Soma: Textual-linguistic and archaeological evidence," *The Indo-Aryans of Ancient South Asia: Language, Material Culture and Ethnicity*, G. Erdosy (ed.), Berlin, Walter de Gruyter & Co., 1995, pp. 370-371.

[6] F. Max Müller, *Biographies of Words and The Home of the Aryas*, London, Longmans, Green, and Co., 1888, p. 235.

[7] Rájendralála Mitra, "Spirituous Drinks in Ancient India," *Journal of the Royal Asiatic Society of Bengal*, 42 (1873), pp. 2, 16, 21.

[8] Abel Bergaigne, *Vedic Religion*, V. G. Paranjpe (trans.), Delhi, Motilal Banarsidass, 1978, vol. I, pp. x, 150.

[9] Adolf Kaegi, *The Rigveda: The Oldest Literature of the Indians*, R. Arrowsmith (trans.), Boston, Ginn and Company, 1902, p. 72.

[10] Griffith, *Rgveda*, vol. I, p. 251 n. 9.

[11] Arthur Anthony Macdonell, *A Vedic Reader for Students*, Oxford, Clarendon Press, 1917, p. 155; Macdonell (trans.), *Hymns from the Rigveda*, London, Oxford University Press, 1922, pp. 77, 79; Macdonell, *Vedic Mythology*, pp. 105, 114.

[12] Arthur Berriedale Keith (trans.), *Rigveda Brahmanas: The Aitareya and Kausītaki Brāhmanas of the Rigveda*, Cambridge, Harvard University Press, 1920, p. 512 n. 4; Keith, *The Religion and Philosophy of the Veda and Upanishads*, Cambridge, Harvard University Press, 1925, vol. II, p. 624.

Washburn Hopkins[1] all shared this view. Georges Dumézil also hypothesized that the Indo-European "drink of immortality"[2] was originally a kind of beer derived from barley.[3]

In *The Road to Eleusis*, R. Gordon Wasson, Albert Hofmann, and Carl A. P. Ruck proposed that the initiates of the Eleusinian Mysteries consumed a psychoactive extract of ergot (*Claviceps* spp.), a fungus parasitic to cultivated grains and wild grasses, and their thesis provides the theoretical basis for this monograph.[4] The year of publication Wasson wrote to Hofmann: "I foresee that our *Eleusis* may lead to a challenge of my identity for Soma in *SOMA: Divine Mushroom of Immortality*. The Aryans of the RigVeda cultivated both barley and wheat. The crucial test might well be this: when the dried sclerotia of *Claviceps purpurea* is reflated with water and poured out, as was the Aryans' practice, what is the color of the resulting fluid?"[5] Wasson would later write to Hoffman: "the ergot possibility for Soma naturally occurred to me. This must be thoroughly explored. I have an open mind. There are difficulties in the elimination of *A. muscaria*. The color suggestions in the RgVeda are too numerous perhaps to be dismissed lightly."[6]

Outlining the complexities of the Soma enigma and the diverse academic disciplines to which this question in the history of science belongs, Richard Evans Schultes writes:

> The Soma question does not interest chemists and pharmacologists alone. It impinges on botany and especially on that field of botany which concerns itself with psychotropic plants. It bears on anthropology and archæology, on religion and the early cultural history of Eurasia. Paraphrasing Georges Clemenceau on

[1] E. Washburn Hopkins, "Soma," *Encyclopædia of Religion and Ethics*, J. Hastings (ed.), Edinburgh, T. & T. Clark, 1920, vol. XI, pp. 685-687.

[2] Gilbert Slater also advanced the hypotheses that Amrita was either Egyptian beer or the fermented juice of the date palm, the date palm theory strongly refuted by Keith. Gilbert Slater, *The Dravidian Element in Indian Culture*, London, Ernest Benn Limited, 1924, p. 79; Keith, *Religion and Philosophy of the Veda*, vol. II, p. 624.

[3] Georges Dumézil, "Le Festin d'Immortalité: Étude de Mythologie Comparée Indo-Européenne," *Annales du Musée Guimet*, 34 (1924), pp. 279-285.

[4] R. Gordon Wasson, Albert Hofmann and Carl A. P. Ruck, *The Road to Eleusis: Unveiling the Secret of the Mysteries*, New York, Harcourt Brace Jovanovich, Inc., 1978.

[5] Wasson to Hofmann, 19 July 1978; in Thomas J. Riedlinger, "Wasson's Alternative Candidates for Soma," *Journal of Psychoactive Drugs*, 25 (1993), p. 154.

[6] Wasson to Hofmann, 4 November 1978; in Ibid.

war and the generals, I would say that Soma and the *Ṛg Veda* are too important to be left to the Vedists.[1]

[1] Richard Evans Schultes, Foreword to R. Gordon Wasson, *Soma and the Fly-Agaric: Mr. Wasson's Rejoinder to Professor Brough*, Cambridge, Botanical Museum of Harvard University, 1972, p. 8.

Chapter 3: Ergot and Barley

Barley and wheat are the universal cereals of Old World plant domestication and were the founder crops of Neolithic agriculture.[1] Barley was a universal companion of wheat, but in comparison was considered to be an inferior staple; it was the main cereal used for fermentation in the Old World.[2] The knowledge of fermentation is ancient, and the technique involved was discovered long before the Aryan migration into India.[3] On the subject of the technology and antiquity of fermentation, R. J. Forbes writes:

> Fermentation needs fire and pottery. Though simple forms of pottery were known by the end of the Upper Palaeolithic the plants collected in those days gave sweet infusions seldom suitable for fermentation, as in general they contain too little sugar. The techniques of fermenting came with organized agriculture, some traces of which go back to the Upper Palaeolithic (Post Glacial

[1] J. M. Renfrew, "The archaeological evidence for the domestication of plants: methods and problems," *The domestication and exploitation of plants and animals*, P. J. Ucko and G. W. Dimbleby (eds.), Chicago, Aldine Publishing Company, 1969, pp. 149-172; Daniel Zohary and Maria Hopf, *Domestication of Plants in the Old World: The origin and spread of cultivated plants in West Asia, Europe and the Nile Valley*, Third Edition, Oxford, Oxford University Press, 2000, pp. 19-69.

[2] Zohary and Hopf, *op. cit.*, p. 59.

[3] Albert Neuburger, *The Technical Arts and Sciences of the Ancients*, H. L. Brose (trans.), New York, Macmillan Company, 1930, pp. 100-109; Gaetano Forni, "The Origin of Grape Wine: A Problem of Historical-Ecological Anthropology," *Gastronomy: The Anthropology of Food and Food Habits*, M. L. Arnott (ed.), The Hague, Mouton Publishers, 1975, p. 76.

> period, Maz d'Azil, France). However a more regular production of cereals suitable for fermentation came in Neolithic times only. Even the wild species of grapes, berries and other fruit collected and eaten in those days would hardly make a suitable base material for fermentation...Mead, the fermented drink made from honey (and grain) may even be older than agriculture, but this we will probably never know...Nectar and ambrosia were probably mead-like concoctions.[1]

Barley and wheat are susceptible to parasitic infections of ergot, which can also develop on rice and millet.[2] Ergot (*Claviceps purpurea*) commonly infects *Lolium* spp. of wild grasses, including *L. temulentum*, a common weed among cereal crops in Eurasia.[3] Ergot (*Claviceps paspali*) also infects *Paspalum scrobiculatum*, a cultivated grain in India that also grows wild. Sheldon Aaronson writes:

> *P. scrobiculatum* grain, a type of millet, was used as food by large numbers of the poorer classes of people in India recently and for millennia in the past (Lisboa, 1896; [etc.]). Two types of grains have been described from *P. scrobiculatum*: small and pale grain...which is wholesome; larger and darker grain...which is unwholesome (Lisboa, 1896)...In 1946, during a rice shortage, people had to take part of their ration in *P. scrobiculatum* and some of them developed symptoms of tremors, giddiness, excessive perspiration and inability to eat or swallow within 20 min of ingestion; there were no fatalities, and symptoms disappeared within 24 h (Ayyar and Narayanaswamy, 1949). These symptoms are similar to those of small amounts (0.5-2 μg/kg) of lysergic acid diethylamide (LSD) (Gilman et al., 1980)... The unwholesome grain's toxicity has been attributed to infection by the ergot, *Claviceps paspali* Stev. and Hall (Bor, 1960) which infects only *Paspalum* spp. (Mantel, 1975)...Vedic and other ancient literature mention *P. scrobiculatum* (*Kodrava*): *Puranas* (ca. 2000

[1] R. J. Forbes, *Studies in Ancient Technology*, Leiden, E. J. Brill, 1955, vol. III, p. 60.

[2] Frank James Bove, *The Story of Ergot: For Physicians, Pharmacists, Nurses, Biochemists, Biologists and Others Interested in the Life Sciences*, Basel, S. Karger, 1970, pp. 43-44, 47, 58-59.

[3] Ibid., p. 43; Hofmann, *Road to Eleusis*, pp. 33-34; LeRoy G. Holm *et al.*, *The World's Worst Weeds: Distribution and Biology*, Honolulu, University Press of Hawaii, 1977, pp. 314-319.

> Before Present/BP) *Kautilya* (approx. 2400 BP) and *Amara-Kas'a* (approx 1700 BP) (Roy, 1933). *Kodrava* was mentioned in the *Brhattrayi* as was the unwholesome type *madana kodrova* (Singh and Chunekar, 1972). Sushruta (ca. 2000 BP) prescribed *kodrava* in combination with other drugs for the treatment of scorpion sting... *Kodrava* was eaten as a cereal food in India for at least the past four millennia. Fungal diseases of cereals like mildews, rusts and smuts are mentioned in the Vedas (approx. 3200 B.P.) (Butler and Bisby, 1960). According to Caius (1935, 1986), ancient Ayurvedic writers described *P. scrobiculatum* seeds as "sweetish and bitter, tonic, and antidotal to poisons, useful in the treatment of ulcers; it caused constipation and flatulence, upset the physiological balance of the body and led to hallucinations and dysuria." It is likely that most, if not all of these symptoms, came from the fungus rather than the cereal. Chauduri and Pal (1978) stated that "It is said the monks ate the (*P. scrobiculatum*) grains with husk which showed the symptoms of intoxication and they could not stand up. It is also said that this effect lasts for some days".[1]

Ergot is a hard-bodied fungus, its growth accompanied by the development of a non-toxic, sweet, yellowish, mucilaginous substance called honeydew, which also forms on the head of the grain stalk.[2] The ergot fungus produces toxic alkaloids, including ergotamine, ergocristine, ergosine, ergocornine, ergokryptine, and ergonovine.[3] The only alkaloid occurring naturally in ergot with potential hallucinogenic activity is d-lysergic acid N-(1-hydroxyethyl) amide, more simply known as lysergic acid hydroxyethylamide.[4] When ergot is boiled for a number of hours in an acidic solution (*e.g.*, with barley) the toxic alkaloids convert to their pharmacologically inactive isomers, and

[1] Sheldon Aaronson, "*Paspalum* spp. and *Claviceps paspali* in Ancient and Modern India," *Journal of Ethnopharmacology*, 24 (1988), pp. 345-347.

[2] Bové, *op. cit.*, pp. 17-23.

[3] Ibid., pp. 172-185; Klaus Lorenz, "Ergot in Cereal Grains," *CRC Critical Reviews in Food Science and Nutrition*, 11 (1979), p. 317; P. M. Scott *et al.*, "Ergot Alkaloids in Grain Foods Sold in Canada," *Journal of AOAC International*, 75 (1992), pp. 773-779.

[4] Richard Evans Schultes and Albert Hofmann, *The Botany and Chemistry of Hallucinogens*, Springfield, Ill., Charles C. Thomas, 1973, pp. 153-154; A. Fanchamps, "Some Compounds with Hallucinogenic Activity," *Ergot Alkaloids and Related Compounds*, B. Berde and H. O. Schild (eds.), Berlin, Springer-Verlag, 1978, pp. 567, 594-596.

following this process of epimerization the psychoactive alkaloid, isolysergic acid hydroxyethylamide can then be extracted from the detoxified ergot with alcohol. The scientific evidence supporting this method is the subject matter of the following paper.

Migrating into India from Central Asia, the Aryans and their Soma-drinking priests encountered the inhabitants of the Harappan Civilization.[1] This was a considerable agricultural population settled in the Indus Valley that subsisted on cultivated barley and wheat.[2] There is no mention of wheat or rice is the *Ṛgveda*, but there are frequent references to both in the Vedic literature that followed.[3] Barley was the most important cereal in the Ṛgvedic period.[4] Parched barley was eaten whole with Soma juice, and was ground into meal and mixed with milk, butter, curd, water, and Soma juice.[5] In the *Śatapatha-Brāhmaṇa*, barley is declared to have alone remained faithful to the gods in their demon contests:

> Now, the gods and the Asurus, both of them sprung from Pragâpati, were contending. Then all the plants went away from the gods, but the barley plants alone went not from them. The gods then prevailed: by means of these (barley-grains) they attracted to themselves all the plants of their enemies; and because they attracted (yu) therewith, therefore they are called yava (barley). They said, 'Come, let us put into the barley whatever sap there is of all plants!' And, accordingly, whatever sap there was of all plants, that they put into the barley…[6]

[1] T. Burrow, "The Early Āryans," *A Cultural History of India*, A. L. Basham (ed.), Oxford, Clarendon Press, 1975, pp. 20-29.

[2] Sir John Marshall (ed.), *Mohenjo-daro and the Indus Civilization…*, London, Arthur Probsthain, 1931, vol. I, p. 27; Sir Mortimer Wheeler, *The Indus Civilization; Supplementary Volume I*, Cambridge, Cambridge University Press, 1953, p. 62; Madho Sarup Vats, *Excavations at Harappā…*, Varanasi, Bhartiya Publishing House, 1974, vol. I, p. 6; B. B. Lal, "The Indus Civilization," *Cultural History of India*, p. 17.

[3] Arthur Anthony Macdonell and Arthur Berriedale Keith, *Vedic Index of Names and Subjects*, Volumes I-II, London, John Murray & Co., 1912, vol. II, p. 345; Om Prakash, *Food and Drinks in Ancient India (From Earliest Times to c. 1200 A.D.)*, Delhi, Munshi Ram Manohar Lal, 1961, pp. 9-11.

[4] Prakash, *op. cit.*, pp. 7-9.

[5] Ibid., pp. 8-9.

[6] *Śatapatha-Brāhmaṇa* III.6.1.8-10; in Julius Eggeling (trans.), *The Satapatha-Brâhmana: According to the Text of the Mâdhyandina School*, Volumes I-V, Oxford, Clarendon Press, 1882-1900, vol. II, p. 142.

The text also contains the myth of the creation of barley from the tear of Soma: "Varu*n*a once struck king Soma in the eye, and it swelled (a*s*vayat): therefrom a horse (a*s*va) sprung…A tear of his fell down: therefrom the barley sprung; whence they say that the barley belongs to Varu*n*a."[1] In the rites connected with the bringing of Soma to the sacrifice, Soma is addressed as Varuṇa, "the wise guardian of the *amṛta*,"[2] and Varuṇa and Mitra are also mystically identified with the pressed Soma juice in the *Ṛgveda*.[3] Barley is the color of Soma,[4] and barley was mixed with Soma; the Soma juice is called "barley-mixed" (*yávāśir-*)[5] in the *Ṛgveda*.[6] The Soma stalks were pressed with barley,[7] and the sweet Soma juice is compared to barley in the *Ṛgveda*: "This juice have we made sweet for thee like barley, blending it with milk."[8]

The cultivation of the Soma-plant is mentioned in the *Harṣacarita*,[9] and according to the *Śatapatha-Brāhmaṇa*, Syāmaka, or cultivated millet,[10] is most like Soma of all the plants.[11] Soma is the "Ruler of the plants,"[12] and in the Rājasūya, the "Consecration of a King," a Soma-sacrifice performed by a Kṣatriya king, the majestic virtues of four plants, all cereals, are symbolically transferred to the royal sacrificer, which are specified in the *Aitareya-Brāhmaṇa*: "The rice is the lordly power of the plants…thus he confers upon him the lordly power. Large rice is the overlordship of the plants…thus he confers upon him overlordship. Panic seeds are the paramount rule of the plants…thus he confers upon him paramount rule. Barley is the leadership of

1 *Śatapatha-Brāhmaṇa* IV.2.1.11; in Ibid., vol. II, p. 281.
2 *Āpastamba-Śrauta-Sūtra* X.30.15; quoted in Alfred Hillebrandt, *Vedic Mythology*, S. R. Sarma (trans.), Delhi, Motilal Banarsidass, 1980, vol. I, p. 200.
3 *Ṛgveda* IX.77.5; in Griffith, *Rgveda*, vol. II, p. 335.
4 *Atharva-Veda* II.8.3; in William Dwight Whitney (trans.), *Atharva-Veda Samhitā*, Volumes I-II, Cambridge, Harvard University, 1905, vol. I, p. 49.
5 Macdonell and Keith, *op. cit.*, vol. II, p. 188.
6 *Ṛgveda* I.187.9, II.22.1, III.42.7, VIII.92.4; in Barend A. Van Nooten and Gary B. Holland (eds.), *Rig Veda: A Metrically Restored Text with an Introduction and Notes*, Cambridge, Department of Sanskrit and Indian Studies, Harvard University, 1994, pp. 111, 125, 159, 409.
7 *Ṛgveda* IX.68.4; in Griffith, *Rgveda*, vol. II, p. 324.
8 *Ṛgveda* VIII.2.3; in Ibid., vol. II, p. 107.
9 *Harṣacarita* II, p. 44; cited in Prakash, *op. cit.*, p. 186. This is a comparatively late work composed by Bāṇa in the seventh century A.D.
10 Macdonell and Keith, *op. cit.*, vol. II, p. 399.
11 *Śatapatha-Brāhmaṇa* V.3.3.4; in Eggeling, *op. cit.*, vol. III, p. 70.
12 *Ṛgveda* IX.114.2; in Griffith, *Rgveda*, vol. II, p. 382.

the plants...thus he confers upon him leadership."[1]

In the *Atharva-Veda*, rice and barley are praised as medicines, and as deliverers from misfortune: "Rice and barley shall be auspicious to thee... They two drive away disease, they two release from calamity;"[2] and, "We speak to the five kingdoms of the plants with soma the most excellent among them. The darbha-grass, hemp, and mighty barley: they shall deliver us from calamity !"[3] Barley is also identified as the panacea in the same text: "This barley they did plough vigorously...With it I drive off to a far distance the ailment from thy body...The waters verily are healing, the waters chase away disease, the waters cure all (disease): may they prepare a remedy for thee !"[4]

The *Atharva-Veda* also speaks of the gods plowing "barley, combined with honey,"[5] and, "This herb, born of honey, dripping honey, sweet as honey, honied, is the remedy for injuries."[6] In the same text, honey is further connected with the barley-god Varuṇa, the "guardian of the *amṛta*:" "Hail be, ye wise Mitra and Varuṇa: with honey swell ye our kingdom."[7] The text also connects honey with Amrita, the nectar of immortality: "These waters, truly, do support Agni and Soma. May the readily flowing, strong sap of the honey-dripping (waters) come to me, together with life's breath and luster !...When, ye golden-coloured, I have refreshed myself with you, then I ween, ambrosia (amṛ́ta) am I tasting !"[8] The text further states: "*sóma* king of plants, immortal oblation—rice and barley (are) remedial, immortal sons of heaven...Of this *amṛ́ta* we make this man to drink the strength; now do I make a remedy, that he may be one of a hundred years."[9] Barley is praised as a "god" in the *Atharva-Veda*,[10] and without ambiguity the text declares: "The grains of rice,

[1] *Aitareya-Brāhmaṇa* VIII.16; in Keith, *Rigveda Brahmanas*, p. 332.

[2] *Atharva-Veda* VIII.2.18; in Maurice Bloomfield (trans.), *Hymns of the Atharva-Veda: Together with Extracts from the Ritual Books and the Commentaries*, Oxford, Clarendon Press, 1897, p. 57.

[3] *Atharva-Veda* XI.6.15; in Ibid., p. 162.

[4] *Atharva-Veda* VI.91.1-3; in Ibid., pp. 40-41.

[5] *Atharva-Veda* VI.30.1; in Whitney, *op. cit.*, vol. I, p. 302.

[6] *Atharva-Veda* VII.56.2; in Bloomfield, *op. cit.*, p. 29.

[7] *Atharva-Veda* VI.97.2; in Ibid., p. 122.

[8] *Atharva-Veda* III.13.5-6; in Ibid., pp. 146-147.

[9] *Atharva-Veda* VIII.7.20-22; in Whitney, *op. cit.*, vol. II, pp. 500-501.

[10] *Atharva-Veda* VI.142.1-2; in Bloomfield, *op. cit.*, p. 141; Ralph T. H. Griffith (trans.), *The Hymns of the Atharva-Veda*, Second Edition, Volumes I-II, Benares, E. J. Lazarus & Co., 1916-1917, vol. I, p. 325.

of barley, that are scattered out—those are soma-shoots."[1]

Soma is called a bull,[2] a tawny bull,[3] and a flying bull,[4] and Soma's "horns" are also mentioned in the *Ṛgveda*.[5] In a hymn addressed to Soma, the text states: "The fearful Bull is bellowing with violent might, far-sighted, sharpening his yellow-coloured horns."[6] The brewed drink of Indra, called Mantha, is also likened to a sharp-horned bull,[7] and Mantha was usually prepared from barley.[8] Soma is implored to grant an abundant harvest: "Pour on us with thy juice all kinds of corn,"[9] and in a hymn addressed to the Pressing-stones that crush the Soma-plant, the *Ṛgveda* states:

> Loudly they speak, for they have found the savoury meath (honey): they make a humming sound over the meat prepared.
>
> As they devour the branch of the Red-coloured Tree, these, the well-pastured Bulls, have uttered bellowings…
>
> Like strong ones drawing, they have put forth all their strength: the Bulls, harnessed together, bear the chariot-poles.
>
> When they have bellowed, panting, swallowing their food, the sound of their loud snorting is like that of steeds…
>
> They have been first to drink the flowing Soma juice, first to enjoy the milky fluid of the stalk.
>
> These Soma-eaters kiss Indra's Bay-coloured Steeds: draining the stalk they sit upon the ox's hide.
>
> Indra, when he hath drunk Soma-meath drawn by them, waxes in strength, is famed, is mighty as a Bull…
>
> This, this the Stones proclaim, what time they are disjoined, and when with ringing sounds they move and drink the balm.
>
> Like tillers of the ground when they are sowing seed, they mix the Soma…[10]

[1] *Atharva-Veda* IX.6.14; in Whitney, *op. cit.*, vol. II, p. 540.
[2] *Ṛgveda* VI.44.21, IX.15.4, IX.70.7; in Griffith, *Rgveda*, vol. I, p. 603, vol. II, pp. 281, 327.
[3] *Ṛgveda* IX.2.9, IX.82.1; in Ibid., vol. II, pp. 270, 338.
[4] *Ṛgveda* IX.86.43; in Ibid., vol. II, p. 347.
[5] *Ṛgveda* IX.5.2, IX.15.4, IX.97.9; in Ibid., vol. II, pp. 272, 281, 359.
[6] *Ṛgveda* IX.70.7; in Ibid., vol. II, p. 327.
[7] *Ṛgveda* X.86.15; in Ibid., vol. II, p. 508; Van Nooten and Holland, *op. cit.*, p. 527.
[8] Macdonell and Keith, *op. cit.*, vol. II, p. 131.
[9] *Ṛgveda* IX.55.1; in Griffith, *Rgveda*, vol. II, p. 303.
[10] *Ṛgveda* X.94.3-13; in Ibid., vol. II, pp. 526-527.

Soma's epithet, "Progenitor of inspired thoughts,"[1] is also used to address the sacrificial cakes of rice or barley flour that are offered to the Vedic gods in the Soma rite, accompanied by oblations of unpounded and boiled rice or barley called *caru*.[2] The cake and *caru* are said to spring from the bull, the father of the sacrifice, the mother being the milch-cow.[3] The *Śatapatha-Brāhmaṇa* states:

> For whatever sap there had been in him (Soma), that sap of his he has produced (extracted) for the offerings…the cake is sap… Thus he unites him with that sap, and so produces him from it,—he (Soma), even when produced, produces him (the sacrificer): hence there is a cake on one potsherd for Varu*n*a. Having made an 'underlayer' of ghee (in the offering-spoon), he says, while making the cuttings from the cake, 'Recite (the invitatory prayer) to Varu*n*a!' Here now some make two cuttings from the Soma-husks, but let him not do so; for that (heap of husks) is an empty body, unfit for offering.[4]

[1] *Ṛgveda* IX.96.5; in Van Nooten and Holland, *op. cit.*, p. 465.
[2] Jan Gonda, *Rice and Barley Offerings in the Veda*, Leiden, E. J. Brill, 1987, pp. 1-6, 14-15, 149-189.
[3] Ibid., p. 15 n. 75.
[4] *Śatapatha-Brāhmaṇa* IV.4.5.15-16; in Eggeling, *op. cit.*, vol. II, p. 382.

Chapter 4: The Preparation of Soma

The original Soma-plant was continuously utilized in the Vedic rites throughout the Brāhmaṇa period and the Śrauta-Sūtra period, which places the extended use of the genuine Soma-plant well into the first millennium B.C.[1] The *Śatapatha-Brāhmaṇa* prescribes various unidentified plants as substitutes for those robbed of the Soma-plant:

> If the Soma is carried off...there are two kinds of Phâlguna plants, the red-flowering and the brown-flowering. Those Phâlguna plants which have brown flowers one may press; for they, the brown flowering Phâlgunas, are akin to the Soma-plant...If they cannot get brown-flowering (Phâlgunas), he may press the *S*yenah*r*ita plant... If they cannot get the *S*yenah*r*ita, he may press Âdâra plants...If they cannot get Âdâras, he may press brown Dûb (dûrvâ) plants, for they, the brown Dûb plants, are akin to the Soma...If they cannot get brown Dûb plants, he may also press any kind of yellow Ku*s*a plants...So much then as to those robbed of their Soma.[2]

According to the *Jaiminīya-Brāhmaṇa*, any plant can be substituted for Soma: "If they should not find this (Soma), they may press out whatever plants there are...all plants are related to Soma."[3] The *Pañcaviṃśa-Brāhmaṇa*

[1] Hillebrandt, *op. cit.*, vol. I, pp. 132-134; C. G. Kashikar, *Identification of Soma*, Pune, C. G. Kashikar, 1990, pp. 9-14.

[2] *Śatapatha-Brāhmaṇa* IV.5.10.1-6; in Eggeling, *op. cit.*, vol. II, pp. 421-422.

[3] *Jaiminīya-Brāhmaṇa* I.355; in H. W. Bodewitz (trans.), *The Jyotiṣṭoma Ritual: Jaiminīya Brāhmaṇa I, 66-364*, Leiden, E. J. Brill, 1990, p. 203.

instructs: "If they (*i.e.*, some rivals) take away the soma before it has been bought, other soma must be bought. If they take it away after it has been bought, other soma, which is to be found in the vicinity, must be obtained; but he should give something (some fee) to the soma-buyer. If they cannot obtain any soma, they should press (instead of soma) pūtīkaplants."[1] The *Mānava-Śrauta-Sūtra* states: "If king soma is stolen, he shall press out the soma that is nearest at hand. He shall give something to the soma seller. If he does not find soma, he shall press out pūtīka herbs."[2] The *Kātyāyana-Śrauta-Sūtra* advises: "If the Soma stalks are stolen away he (the Adhvaryu) should ask the servant of the Sacrificer to run and search…If the Soma-stalks are not found he (the Adhvaryu) should press the Arjuna plants (grass) with ruddy brown flowers."[3] The *Āśvalāyana-Śrauta-Sūtra* prescribes the following: "If the King Soma, after its purchase, is lost or burnt away…they should press the stalk of Pūtīkā and Phālguna plants…If the Soma plant becomes available, he should perform the Soma sacrifice in the normal manner."[4]

The extent of the popularity of Soma in the Ṛgvedic period can be inferred from very limited evidence. A few passages from the text suggest that Soma was drunk for extended periods during sacrificial occasions, and was also consumed in private homes. The *Ṛgveda* refers to the pressing of Soma, "If of a truth in every house, O Mortar, thou art set for work;"[5] and, "O Gods, with constant draught of milk, husband and wife with one accord Press out and wash the Soma juice."[6] The *Ṛgveda* also tells of great Soma pressings, a thousand in number, that is, a yearlong sacrifice, performed by King Bhâvya on the banks of the Sindhu (Indus),[7] and speaks of "These Brâhmans with the Soma juice, performing their year-long rites."[8] The pressed Soma juice

[1] *Pañcaviṃśa-Brāhmaṇa* IX.5.1-3; in W. Caland (trans.), *Pañcaviṃśa-Brāhmaṇa: The Brāhmaṇa of Twenty Five Chapters*, Calcutta, Asiatic Society of Bengal, 1931, p. 212.

[2] *Mānava-Śrauta-Sūtra* III.6.3-4; in Jeannette M. van Gelder (trans.), *The Mānava Śrautasūtra: belonging to the Maitrāyaṇī Saṃhitā*, New Delhi, International Academy of Indian Culture, 1963, p. 114.

[3] *Kātyāyana-Śrauta-Sūtra* XXV.12.17-18; in H. G. Ranade (trans.), *Kātyāyana Śrauta Sūtra: Rules for the Vedic Sacrifices*, Pune, H. G. Ranade and R. H. Ranade, 1978, p. 636.

[4] *Āśvalāyana-Śrauta-Sūtra* VI.8.1-16; in H. G. Ranade (trans.), *Āśvalāyana Śrauta-Sūtram*, Poona, R. H. Ranade, 1981, vol. I, pp. 184-186.

[5] *Ṛgveda* I.28.5; in Griffith, *Ṛgveda*, vol. I, p. 36.

[6] *Ṛgveda* VIII.31.5; in Ibid., vol. II, p. 168.

[7] *Ṛgveda* I.126.1; in Ibid., vol. I, p. 174.

[8] *Ṛgveda* VII.103.8; in Ibid., vol. II, p. 97.

was collected in large wooden tubs,[1] and was produced in considerable quantities: "even as rivers to the ocean, flow forth from days of old the Soma juices."[2] The ceremonial use of Soma continued in the Śrauta ritual during the Brāhmana period and the Śrauta-Sūtra period, although it is not mentioned in the domestic rites, perhaps suggesting that Soma was no longer a common drink.[3] From this era, the *Śāṅkhāyana-Śrauta-Sūtra* tells of Soma-feasts attended by a hundred priests in the performance of the Rājasūya, a Soma-sacrifice consecrating a Kṣatriya king.[4]

The *Ṛgveda* does not contain a systematic, scientific account of the preparation of Soma, and the exact method employed by the Vedic priests has thus far remained concealed in what Max Müller calls the "impregnable fortress of ancient Vedic literature."[5] In the course of the rite, Soma could be pressed three times a day. Referring to the Soma hymns of the *Ṛgveda*, Arthur Anthony Macdonell writes: "The whole of the ninth book consists of incantations chanted over the tangible Soma, while the stalks are being pounded by stones, the juice passes through a woollen strainer, and flows into wood vats, in which it is offered to the gods on the litter of sacred grass (barhís).[6] These processes are overlaid with confused and mystical imagery in endless variation."[7] Rájendralála Mitra writes:

> The soma nectar…for aught we know, was never manufactured for sale; but it was associated with the earliest history of the Aryans, even before they separated from the ancient Persians, and enjoyed the proud pre-eminence of a god…The Rig Veda Sañhitá is most lavish in its praise, and all the four Vedas furnish innumerable mantras for repetition at every stage of its manufacture…nothing could be done without appropriate mantras, and the ritual throughout

1 *Ṛgveda* IX.15.7, IX.33.2; in Ibid., vol. II, pp. 281, 292; B. H. Kapadia, *A Critical Interpretation and Investigation of Epithets of Soma*, Vallabh Vidyanagar, B. H. Kapadia, 1959, pp. 12-13, 16.

2 *Ṛgveda* III.46.4; in Griffith, *Ṛgveda*, vol. I, p. 367.

3 Prakash, *op. cit.*, p. 44.

4 *Śāṅkhāyana-Śrauta-Sūtra* XV.14.9; in W. Caland (trans.), *Śāṅkhāyana-Śrautasūtra: being a major yājñika text of the Ṛgveda*, Nagpur, International Academy of Indian Culture, 1953, p. 436.

5 F. Max Müller (trans.), *Vedic Hymns*, Oxford, Clarendon Press, 1891, vol. I, p. xxviii.

6 The barhís is a seat of darbha-grass, which is also added to the Soma beverage.

7 Macdonell, *Vedic Reader*, p. 153.

was most complicated and tedious.[1]

The *Aitareya-Brāhmaṇa* prescribes the following in the performance of the Rājasūya, a Soma-sacrifice consecrating a Kṣatriya king:

> These things the Adhvaryu should make ready in advance; the skin for pressing, the two pressing boards, the wooden tub, the filter cloth, the pressing stones, the vessel for the pure Soma, the stirring vessel, the vessel, the drawing cup, and the goblet. When they press the king in the morning, then he should divide these (fruits) in two; some he should press, the rest leave over for the midday pressing. When they fill up the goblets, then he should fill up the goblet of the sacrificer; in it should have been cast two Darbha shoots...When they lift them up (to the mouth), then he should lift it up after them. When the Hotṛ invokes the sacrificial food, when he partakes of the food in the goblet, then he should partake of it with
>
> "That which is left over of the pressed juice rich in sap,
> Which Indra drank mightily,
> Here with auspicious mind this of him,
> I partake of Soma the king."
>
> Auspiciously to him this (food) from the trees is consumed with auspicious mind, dread is his sway, unassailable, who as a Kṣatriya when sacrificing partakes thus of this food. With
>
> "Be thou kindly to our heart when drunk,
> Do thou extend our life, to live long, O Soma"...[2]

The Adhvaryus were the Vedic priests who prepared the Soma,[3] and the *Ṛgveda* speaks of the Adhvaryus "sweating with their kettles."[4] In the preparation of the Soma beverage, the Soma stalks were boiled,[5] and remained standing for three nights.[6] The *Taittirīya-Saṃhitā* contains the mythological justification for keeping Soma for three nights: "When the Soma was being

[1] Mitra, *op. cit.*, p. 21.
[2] *Aitareya-Brāhmaṇa* VII.32-33; in Keith, *Rigveda Brahmanas*, pp. 316-317.
[3] *Ṛgveda* V.43.3-5; in Griffith, *Ṛgveda*, vol. I, p. 508; *Aitareya-Brāhmaṇa* II.20; in Keith, *Rigveda Brahmanas*, p. 149.
[4] *Ṛgveda* VII.103.8; in Griffith, *Rgveda*, vol. II, p. 97.
[5] *Śatapatha-Brāhmaṇa* XII.7.3.6; in Eggeling, *op. cit.*, vol. V, p. 224.
[6] *Baudhāyana-Śrauta-Sūtra* XVII.32; in R. N. Dandekar (trans.), *Śrautakośa: Encyclopædia of Vedic Sacrificial Ritual Comprising the Two Complementary Sections, Namely, the Sanskrit Section and the English Section*, Poona, Vaidika Saṁsodhana Maṇḍala, 1962, vol. I, pt. 2, p. 904.

borne away, the Gandharva Viçvāvasu stole it. It was for three nights stolen; therefore after purchase the Soma is kept for three nights."[1] Referring to Surā, an alcoholic beverage produced from fermented grain, the *Śatapatha-Brāhmaṇa* states: "He distils (i.e., boils) it with a view to (its being like) the Soma-pressing. For three nights it remains standing, for the Soma remains standing for three nights after it has been bought."[2] The *Atharva-Veda* also makes reference to "him that takes pains, and cooks and presses the soma."[3] According to the Vedic literature, the Soma liquor was fermented. The directions in the Soma rite include the word Parisrut, "spirituous liquor,"[4] and this is also indicated in the *Ṛgveda* by the term Vâtâpi, "fermenting" of the Soma liquor:[5] "What, Soma, we enjoy from thee in milky food or barley-brew, Vâtâpi, grow thou fat thereby."[6]

Following the fermenting of the liquor, the Soma stalks were removed from the vessel and crushed between two stones on a cowhide,[7] or with a mortar and pestle.[8] The *Ṛgveda* tells of Soma libations "fifteenfold strong,"[9] and with the addition of water,[10] which diluted the mixture and caused the crushed Soma stalks to swell,[11] the Soma juice was poured over the stalks upon the cowhide, and filtered through a strainer made of cloth or wool.[12] The Soma stalks were brown (*babhru*), ruddy (*aruṇa*), or tawny (*hari*)[13] in

[1] *Taittirīya-Saṃhitā* VI.1.6.4; in Arthur Berriedale Keith (trans.), *The Veda of the Black Yajus School entitled Taittiriya Sanhita*, Volumes I-II, Cambridge, Harvard University Press, 1914, vol. II, p. 493.

[2] *Śatapatha-Brāhmaṇa* XII.7.3.6; in Eggeling, *op. cit.*, vol. V, p. 224.

[3] *Atharva-Veda* XI.1.30; in Bloomfield, *op. cit.*, p. 184.

[4] *Śatapatha-Brāhmaṇa* XII.8.2.15; in Eggeling, *op. cit.*, vol. V, p. 243.

[5] *Ṛgveda* I.187.10; in Griffith, *Rgveda*, vol. I, p. 251 n. 9; *Maitrāyaṇī-Saṃhitā* I.9.1; cited in Keith, *Rigveda Brahmanas*, p. 512 n. 4.

[6] *Ṛgveda* I.187.10; in Griffith, *Rgveda*, vol. I, p. 251.

[7] *Ṛgveda* IX.65.25, IX.79.4, X.94.9; in Griffith, *Rgveda*, vol. II, pp. 317, 336, 527.

[8] *Ṛgveda* I.28.1-6, IX.46.3; in Ibid., vol. I, pp. 36-37, vol. II, p. 299.

[9] *Ṛgveda* X.27.2; in Ibid., vol. II, p. 416.

[10] *Ṛgveda* IX.75.9, IX.107.2; in Ibid., vol. II, pp. 333, 373; Hillebrandt, *op. cit.*, vol. I, pp. 301-305.

[11] *Ṛgveda* VIII.9.19, IX.64.8, IX.107.12; in Griffith, *Rgveda*, vol. II, pp. 130, 314, 374; *Taittirīya-Saṃhitā* I.2.11a; in Keith, *Taittiriya Sanhita*, vol. I, pp. 29-30; Macdonell, *Vedic Mythology*, p. 107.

[12] *Ṛgveda* VIII.2.2, IX.12.4, IX.13.1; in Griffith, *Rgveda*, vol. II, pp. 107, 278, 279; B. H. Kapadia, *Soma in the Legends*, Vallabh Vidyanagar, B. H. Kapadia, 1958, p. 32.

[13] "[T]he cap of the fly-agaric is red, while the colour of haoma is given as *zari* (Sanskrit *hari*), which in Avestan firmly means 'green' or 'yellowish green'." Ilya Gershevitch, "An Iranianist's View of the Soma Controversy," *Mémorial Jean De Menasce*, A. Tafozzoli (ed.), Louvain, Imprimerie Orientaliste, 1974, pp. 47, 58-59.

color,[1] and the Soma juice was also brown,[2] ruddy,[3] or tawny.[4] This beverage is described in the *Ṛgveda* as "good to taste and full of sweetness, verily it is strong and rich in flavour."[5] Indicating that the fermenting of the liquor preceded the pressing, Soma was often drunk unmixed,[6] or it was mixed with milk,[7] butter,[8] curd,[9] or barley.[10]

---

[1] *Ṛgveda* VIII.9.19; in Griffith, *Rgveda*, vol. II, p. 130; *Taittirīya-Saṃhitā* VI.1.6.7; in Keith, *Taittiriya Sanhita*, vol. II, p. 494 and n. 3; *Śatapatha-Brāhmaṇa* III.3.1.13-16; in Eggeling, *op. cit.*, vol. II, pp. 62-63; Macdonell, *Vedic Mythology*, p. 105.

[2] *Ṛgveda* IX.33.2, IX.63.4, 6; in Griffith, *Rgveda*, vol. II, pp. 292, 311; Van Nooten and Holland, *op. cit.*, pp. 434, 445.

[3] *Ṛgveda* IX.45.3; in Griffith, *Rgveda*, vol. II, p. 298; Van Nooten and Holland, *op. cit.*, p. 438.

[4] *Ṛgveda* IX.3.9, IX.98.7; in Griffith, *Rgveda*, vol. II, pp. 271, 365; Van Nooten and Holland, *op. cit.*, pp. 421, 468.

[5] *Ṛgveda* VI.47.1; in Griffith, *Rgveda*, vol. I, p. 609.

[6] *Ṛgveda* I.135.3, V.2.3, VII.90.2, IX.72.4; in Ibid., vol. I, pp. 187, 467, vol. II, pp. 86, 329.

[7] *Ṛgveda* I.23.1, VIII.2.3, VIII.90.10, IX.11.2, 5, IX.64.28, IX.72.1, IX.101.12, IX.107.2; in Ibid., vol. I, p. 28, vol. II, pp. 107, 253, 278, 315, 329, 369, 373.

[8] *Ṛgveda* X.29.6; in Ibid., vol. II, p. 423.

[9] *Ṛgveda* I.5.5, V.51.7, VII.32.4, IX.11.6, IX.101.12; in Ibid., vol. I, pp. 7, 518, vol. II, pp. 32, 278, 369.

[10] *Ṛgveda* I.187.9, III.35.3, 7, III.52.1, IX.68.4; in Ibid., vol. I, pp. 251, 356, 371, vol. II, p. 324.

Chapter 5: Soma and Surā in the *Sautrāmaṇī*

The Śrauta ritual comprises two sacrificial institutions: the Soma-sacrifices and the Havis-sacrifices (*i.e.*, food offerings), and the Kaukilī Sautrāmaṇī belongs to the latter.[1] The Sautrāmaṇī-sacrifice has two forms: the Caraka and Kaukilī. The Vedic texts provide detailed recipes and formulas for the preparation of an alcoholic beverage called Surā in both forms of the Sautrāmaṇī, and both forms were performed for similar reasons.[2] The *Vaitāna-Sūtra* instructs: "One should not perform the *Sautrāmaṇī* without having first performed the Soma-sacrifice,"[3] and both forms were prescribed for one who vomited or evacuated Soma.[4] Only the Caraka Sautrāmaṇī is performed at the end of the Rājasūya Soma-sacrifice consecrating a Kṣatriya king.[5] Between the two forms of the Sautrāmaṇī, the procedures recorded in the Vedic texts for the preparation of Surā contains relatively minor differences, and on the surface, the main difference is the addition of milk

---

[1] Dandekar, *op. cit.*, vol. I, pt. 2, pp. 5, 899.

[2] *Mānava-Śrauta-Sūtra* V.2.11.2; in Gelder, *op. cit.*, p. 176; *Vārāha-Śrauta-Sūtra* III.2.8.1; in Dandekar, *op. cit.*, vol. I, pt. 2, p. 928.

[3] *Vaitāna-Sūtra* XXX.1; in Dandekar, *op. cit.*, vol. I, pt. 2, p. 938.

[4] *Mānava-Śrauta-Sūtra* V.2.4.1; in Gelder, *op. cit.*, p. 162; *Kātyāyana-Śrauta-Sūtra* XIX.1.2; in Ranade, *Kātyāyana Śrauta Sūtra*, p. 517; *Baudhāyana-Śrauta-Sūtra* XXIII.16; in Dandekar, *op. cit.*, vol. I, pt. 2, p. 905; *Vārāha-Śrauta-Sūtra* III.2.7.1; in Ibid., vol. I, pt. 2, p. 916.

[5] *Mānava-Śrauta-Sūtra* V.2.4.1; in Gelder, *op. cit.*, p. 162; *Kātyāyana-Śrauta-Sūtra* XV.10.24; in Ranade, *Kātyāyana Śrauta Sūtra*, p. 447; *Āpastamba-Śrauta-Sūtra* XIX.4; in Dandekar, *op. cit.*, vol. I, pt. 2, p. 912; *Satyāṣāḍha-Śrauta-Sūtra* XIII.8.4; in Ibid., vol. I, pt. 2, p. 913; *Vārāha-Śrauta-Sūtra* III.2.7.1; in Ibid., vol. I, pt. 2, p. 916.

to the Surā liquor in the Kaukilī Sautrāmaṇī.[1] The rituals in both forms are also very similar, and the Kaukilī Sautrāmaṇī is an elaboration of the Caraka Sautrāmaṇī, which indicates that the latter is the primary form of the rite.[2] In the Caraka Sautrāmaṇī, the Surā liquor is intimately connected with Soma, and this form of the rite is a Soma-sacrifice.

The Surā liquor of the Sautrāmaṇī-sacrifice, as described in the Vedic texts, is a kind of beer, although this word is often translated as "wine."[3] According to the literature, Surā is prepared with barley, rice, and sometimes wheat.[4] The fermentation of the liquor was accomplished through the use of malted grain, which produces the sugars necessary for the production of alcohol. Malting can be achieved simply by sprinkling the grain with water and leaving it to germinate, protected from the direct sunlight.[5]

> The basis of all beer production is the fermentation of starch in amylaceous cereals. Grain always contains a small quantity of directly fermentable sugar, but this is inadequate in amount to produce an alcoholic drink. As starch, itself, cannot ferment unless first split into fermentable sugars, it is usually first subjected to malting; i.e. letting grain germinate, a process during which starch is converted into maltose, and considerable amounts of diastase are developed. In modern processes, malt is then heated and dried to stop germination, then boiled with water, strained and incubated with yeast.[6]

Soma is mentioned repeatedly with Surā in the Vedic texts, and the two beverages were also combined together.[7] In the Caraka Sautrāmaṇī, a mare is given away as a sacrificial gift, and the reason for this is provided in the *Taittirīya-Brāhmaṇa*: "The mare gives birth to a horse or to a mule. The *sautrāmaṇī* is either a *soma* (sacrifice) or a *surā* (sacrifice)."[8] The text also states that Soma and Surā are a pair: "husband and wife–Soma is the best

---

[1] Madhavi Bhaskar Kolhatkar, *Surā: The Liquor and the Vedic Sacrifice*, New Delhi, D. K. Printworld Ltd., 1999, pp. 119-132.

[2] Dandekar, *op. cit.*, vol. I, pt. 2, pp. 899-902; Kolhatkar, *op. cit.*, pp. 137-163, 181.

[3] Kolhatkar, *op. cit.*, pp. 2, 118, 135.

[4] Prakash, *op. cit.*, pp. 26, 44, 300.

[5] Forbes, *op. cit.*, vol. III, pp. 63-64.

[6] William J. Darby, Paul Ghalioungui and Louis Grivetti, *Food: The Gift of Osiris*, London, Academic Press, 1977, vol. II, p. 534.

[7] Hillebrandt, *op. cit.*, vol. I, pp. 321-324, 468 n. 291.

[8] *Taittirīya-Brāhmaṇa* I.8.6.4; quoted in Kolhatkar, *op. cit.*, p. 112.

food of the gods, Surā the best food of men."[1] According to the *Śatapatha-Brāhmaṇa*, the Sautrāmaṇī is a Soma-sacrifice because the formulas in the rite contain the word Parisrut, "spirituous liquor."[2] The text also declares that the Sautrāmaṇī belongs to the great Soma-drinking god Indra,[3] and through the performance of this rite the sacrificer obtains immortality.[4] The text states: "the Sautrâma*n*î is the same as the Soma (sacrifice)."[5]

The Sautrāmaṇī extends over a four-day period, the first three of which the Surā is prepared and fermented. The *Śāṅkhāyana-Śrauta-Sūtra* instructs that the Kaukilī Sautrāmaṇī is "to be performed as sacrifices of Soma."[6] Julius Eggeling provides a description of this particular rite:

> The whole performance takes four days, during the first three of which the Surâ-liquor is prepared and matured, and offerings of a rice-pap to Aditi, and a bull to Indra are performed; whilst the main sacrifice takes place on the fourth day—the day of either full moon or new moon—the chief oblations offered on that day being three cups of milk, and as many of Surâ-liquor, to the A*s*vins, Sarasvatî, and Indra respectively; of three animal victims to the same deities... At the end of the sacrifice, a third bull is offered to Indra in his form of Vayodhas (giver of life), together with another pap (*k*aru) to Aditi and an oblation of curds to Mitra and Varu*n*a. No mention is made of the Agnishomîya he-goat usually offered on the day preceding the Soma-pressing, the first bull offered to Indra probably taking its place on this occasion, whilst the bull to Indra Vayodhas would seem to take the place of the sacrifice of a barren cow (to Mitra and Varu*n*a) which usually takes place at the end of a Soma-sacrifice.[7]

The filtering of the Surā liquor occurs on the fourth day:

> This performance thus takes place on the fourth day. Behind the mound of the southern Vedi a whole is dug, and an ox-hide spread over it. On this skin the unstrained liquor (parisrut) is either poured, a fine strainer (made of bamboo) being then laid thereon so that

[1] *Taittirīya-Brāhmaṇa* I.3.3.2; quoted in Hillebrandt, *op. cit.*, vol. I, pp. 321-322.
[2] *Śatapatha-Brāhmaṇa* XII.8.2.15; in Eggeling, *op. cit.*, vol. V, p. 243.
[3] *Śatapatha-Brāhmaṇa* XII.8.2.24; in Ibid., vol. V, p. 245.
[4] *Śatapatha-Brāhmaṇa* XII.9.1.7-9; in Ibid., vol. V, pp. 261-262.
[5] *Śatapatha-Brāhmaṇa* XII.9.2.1; in Ibid., vol. V, p. 264.
[6] *Śāṅkhāyana-Śrauta-Sūtra* XIV.13.14; in Caland, *Śāṅkhāyana-Śrautasūtra*, p. 382.
[7] Eggeling, *op. cit.*, vol. V, pp. 213-214 n. 2.

> the clear liquor percolates through the holes, and the dregs remain below; or the strainer is placed on the skin, and the unstrained liquor is poured on it so as to allow the clear liquor to flow through on the skin. The liquor is then poured into a pan (sata), and further purified by a whisk of cow and horse-hair being drawn through it, or the liquor being strained through the hair.[1]

With the exception of the addition of milk to the mixture in the Kaukilī Sautrāmaṇī, the Surā liquor was produced exclusively from cereals,[2] and the Vedic texts prescribe that these ingredients are to be purchased from a vendor of Soma, or a eunuch.[3] During the ritual of purchasing Soma, the Soma stalks were placed on a red ox-hide.[4] The *Baudhāyana-Śrauta-Sūtra* prescribes that the rice and barley stalks for Surā in the Caraka Sautrāmaṇī are to be placed on a red bull-hide during purchase, which are referred to as Surāsoma: "the *surāsoma* (= the tender grass of barley and paddy used for preparing wine) should remain placed upon the red bull's hide...The eunuch should sit down to the south of the *surāsoma*. The adhvaryu should purchase from the eunuch the tender grass in return for lead, saying 'This (= lead) is yours; this (= tender grass) is mine.' (The eunuch should say,) 'The *surāsoma* has been sold out.'"[5] As part of the ritual, the grain stalks purchased for Surā are tied up in cloth,[6] as the Soma stalks were also tied in cloth.[7]

The instructions for the Caraka Sautrāmaṇī in the *Vārāha-Śrauta-Sūtra* state: "One should purchase either from a eunuch or from a (Soma) vendor tender grass in return for lead...He should cook rice over the gārhapatya fire, mix up parched grains, tokma (malted barley), and nagnahu (barley flour) into that rice, and thus prepare the māsara (wine). He should add to it tender grass with the (formulaic) verse...He should allow that mixture to remain for three nights."[8] The *Āpastamba-Śrauta-Sūtra* instructs that Surā in the Caraka Sautrāmaṇī is prepared "according to the normal procedure of preparing

[1] Ibid., vol. V, p. 226 n. 1.
[2] Dandekar, *op. cit.*, vol. I, pt. 2, pp. 899-900; Kolhatkar, *op. cit.*, pp. 137-163, 181.
[3] *Śatapatha-Brāhmaṇa* XII.7.2.12; in Eggeling, *op. cit.*, vol. V, pp. 219-220.
[4] Eggeling, *op. cit.*, vol. II, p. 64 n. 2.
[5] *Baudhāyana-Śrauta-Sūtra* XVII.31-32; in Dandekar, *op. cit.*, vol. I, pt. 2, p. 903.
[6] *Kātyāyana-Śrauta-Sūtra* XV.9.24; in Ranade, *Kātyāyana Śrauta Sūtra*, p. 444.
[7] *Śatapatha-Brāhmaṇa* III.3.2.4; in Eggeling, *op. cit.*, vol. II, p. 64; *Kātyāyana-Śrauta-Sūtra* VII.7.2-3; in Ranade, *Kātyāyana Śrauta Sūtra*, p. 237.
[8] *Vārāha-Śrauta-Sūtra* III.2.7.1; in Dandekar, *op. cit.*, vol. I, pt. 2, p. 916.

wine. This wine is called parisrut. He should put in it the tender sprouts of barley with the formula...and allow that mixture to remain for three nights... he should pass the wine through the filter made of hair in another vessel."[1] In the same text, the recipe for Surā in the Kaukilī Sautrāmaṇī is as follows:

> One should have purchased in advance paddy, barley, and *śyāmāka* [millet]...One should roll up the paddy in a piece of linen cloth and thereby turn it into *tokma* (that is to say, one should allow sprouts to grow on that paddy.) He should parch barley, crush it into flour, put that flour into curds or butter-milk, and stir up that mixture with *darbha*-blades. This mixture is called *māsara*. One should have the thick flour of barley sprinkled with water. This is called *nagnahu*. One should parch *śyāmāka* and crush it into flour. At the time of preparing the wine, he should mix together the *tokma*, the *māsara*, and the *nagnahu*, and scatter upon that mixture one-third of the *śyāmāka*-flour...he should pour out upon it the milk of one cow. (After one night has passed) he should again scatter upon it another third of the *śyāmāka*-flour and add to it the milk of two cows...(After the second night has passed) he should again scatter upon it the remaining third of the *śyāmāka*-flour and add to it the milk of three cows...[2]

The instructions for the Caraka Sautrāmaṇī in the *Kātyāyana-Śrauta-Sūtra* include the following: "The adhvaryu should purchase from a eunuch, in return for lead, paddy which is sprouted and which is not sprouted, and tie it up into a piece of linen cloth. He should cook rice and mix with it the flour of the sprouted paddy, with the (three) formulas...He should perform in respect of the materials for wine the rites which are usually performed in respect of the preparation of a cake."[3] The more elaborate instructions for the Kaukilī Sautrāmaṇī in the same text are as follows:

> One should then make purchases from him (= the vendor)...on a hide (of bull)—tender grass in return for lead, *tokma* in return for wool, and parched paddy in return for threads. With reference to each of the above-mentioned materials (as also with reference

[1] *Āpastamba-Śrauta-Sūtra* XIX.1-2; in Ibid., vol. I, pt. 2, p. 909.
[2] *Āpastamba-Śrauta-Sūtra* XIX.5; in Ibid., vol. I, pt. 2, p. 922.
[3] *Kātyāyana-Śrauta-Sūtra* XV.9.24-28; in Ibid., vol. I, pt. 2, p. 918.

> to *nagnahu*), he should say to the *surāsoma*-vendor: "O Vendor of *surāsoma*, is your *surāsoma* to be sold?" Some teachers say that one should purchase these things from a eunuch. He should carry those materials into the fire-hall by the southern door and pulverize the *nagnahu* and also the other materials. He should cook rice-grains and *śyāmāka*-grains, in such a way that there is amble water, pour out the scum of these two in two vessels, and mix up that scum with flour of *nagnahu*. This is *māsara*. He should then mix both the (potfuls of) cooked grains (poured out into a pitcher) with the pulverized *nagnahu* and *māsara*...He should keep that pitcher for three nights...he should sprinkle the wine with the milk...and scatter upon it the flour of tender grass. Next morning, he should sprinkle the wine with the milk...and scatter upon it the flour of *tokma*. On the third morning, he should sprinkle the wine with the milk...and scatter upon it the flour of parched paddy.[1]

In the mixing of the cereal ingredients for Surā in the Caraka Sautrāmaṇī, the Adhvaryu turns the liquor into Soma with the formulas: "I mix with the Soma," and "Thou art Soma: get thee matured for the Aśvins! get thee matured for Sarasvatî! get thee matured for Indra Sutrâman!"[2] The instructions for the Caraka Sautrāmaṇī in the *Mānava-Śrauta-Sūtra* are as follows:

> One who has purged soma, one who is consecrated as king by the rājasuya, one desirous of prosperity or one who has long been ill shall offer the sautrāmaṇī as a sacrifice (iṣṭi). For lead he buys from a eunuch in a fenced place, as material for the surā, young shoots of rice, barley and wheat for fermenting. Having said: "Seller of surā-soma, for this lead let me buy surā-soma from thee" he buys. With: "Thee the sweet with the sweet, the bitter with the bitter, the bright with the bright, the divine with the divine, the immortal with the immortal, thee with the soma I unite" he prepares (the surā). With "Thou art soma; ripen for the Aśvins, ripen for Sarasvatī, ripen for Indra the good protector (sutrāman)" he touches

[1] *Kātyāyana-Śrauta-Sūtra* XIX.1.18-21; in Ibid., vol. I, pt. 2, pp. 931-932.

[2] *Śatapatha-Brāhmaṇa* XII.7.3.5-6; in Eggeling, *op. cit.*, vol. V, p. 224; *Mānava-Śrauta-Sūtra* V.2.4.4-5; in Gelder, *op. cit.*, p. 162; *Baudhāyana-Śrauta-Sūtra* XVII.32; in Dandekar, *op. cit.*, vol. I, pt. 2, p. 904; *Satyāṣādha-Śrauta-Sūtra* XIII.8.1; in Ibid., vol. I, pt. 2, p. 912; *Vārāha-Śrauta-Sūtra* III.2.7.5; in Ibid., vol. I, pt. 2, p. 916.

> it. (After saying): "Tomorrow the surā will be produced", he stays overnight for the animal sacrifice...The pratiprasthātṛ spreads the hair sieve over the wooden tub and purifies the surā, with: "May the daughter of the sun purify for thee the fermented liquor, the soma, continually with the eternal hair sieve". With the verse: "What then? As men who have barley reap the barley..." the adhvaryu scoops the milk draughts, holding the two strainers of young shoots above them; (the pratiprasthātṛ the surā draughts) for the Aśvins, for Sarasvatī, for Indra...The pratiprasthātṛ mixes the surā draughts in the supporting vessels, that for the Aśvins with ground kuvala fruits...that for Sarasvatī with ground karkandhu fruits, and that for Indra with ground badara fruits...(The adhvaryu) makes (the maitrāvaruṇa) recite with: "Recite the invitatory verse for the gladdening soma draughts to the Aśvins, to Sarasvatī, to Indra the good protector"...After having poured the remains (of the surā and the milk) into the two principal (vāyu vessels) and having brought them round in front of the two fires, the sacrificer enjoys them in the south; with the verse: "What is left here from the juicy, pressed out (soma), of which Indra drank mightily, this, king soma I enjoy here with luminous mind" the milk draughts; with the verse: "For various seats are arranged for you two among the gods..." the surā draughts...[1]

The same text instructs that the procedures in both forms of the Sautrāmaṇī are the same until the preparation of Surā, the mixing of the ingredients with Soma in the Caraka Sautrāmaṇī replaced with milk in the Kaukilī Sautrāmaṇī:[2]

> The ritual for the Kaukilī sautrāmaṇī is the same as that for the sautrāmaṇī as sacrifice. (It is offered) for the same wishes and for one who has vomited soma. The procedure is normal up to the preparation. With the verse: "Pour around here the pressed soma,

[1] *Mānava-Śrauta-Sūtra* V.2.4.1-29; in Gelder, *op. cit.*, pp. 162-163.

[2] The text prescribes that in the Kaukilī Sautrāmaṇī the offerings to the gods are further mixed: "[The pratiprasthātṛ] mixes the (surā draughts) for the Aśvins with ground kuvala fruits, ground wheat and the hair of a wolf, that for Sarasvatī with ground karkandhu fruits, ground Indragrains and the hair of a tiger, that for Indra with ground badara fruits, ground barley and the hair of a lion." *Mānava-Śrauta-Sūtra* V.2.11.16; in Ibid., p. 177.

> the highest oblation; I have pressed out with the stones the strong soma, that has run within the water" he pours fresh milk around (the surā), milk from one cow at the first dawning, from two at the second. (After saying): "In three days the surā will be produced" he stays overnight for the animal sacrifice.[1]

Most texts prescribe that Surā ferment for three nights after it was mixed, although the *Śatapatha-Brāhmaṇa* allows one day.[2] Soma was kept for three nights after it was purchased,[3] and there were also Soma rites that were performed in one day.[4] The *Baudhāyana-Śrauta-Sūtra* instructs that Surā in the Caraka Sautrāmaṇī should remain standing, like Soma, for three nights: "This mixture should remain (like this) for three nights. (For,) it is said in the [*Taittirīya-*] *Brāhmaṇa*: 'The Soma, which is purchased, remains (undisturbed for three nights).'"[5]

The *Śatapatha-Brāhmaṇa* contains the mythology connected with the origin of the Sautrāmaṇī-sacrifice, the Caraka Sautrāmaṇī being the primary form of the rite:

> Now Tvash*tri* had a three-headed, six-eyed son. He had three mouths; and because he was thus shapen, he was called Vi*s*varûpa ('All-shape'). One of his mouths was Soma-drinking, one spirit-drinking, and one for other food. Indra hated him, and cut off those heads of his...Tvashtri was furious: 'Has he really slain my son?' He brought Soma-juice withheld from Indra; and so that Soma-juice was, when produced, even so it remained withheld from Indra. Indra thought within himself: 'There now, they are excluding me from Soma!' and even uninvited he consumed what pure (Soma) there was in the tub, as the stronger (would consume the food) of the weaker. But it hurt him: it flowed in all directions from (the openings of) his vital airs; only from his mouth it did not flow... and what flowed from the upper opening that was the foaming spirit (parisrut). And thrice he spit out: thence were produced the (fruits

[1] *Mānava-Śrauta-Sūtra* V.2.11.1-5; in Ibid., p. 176.
[2] *Śatapatha-Brāhmaṇa* V.5.4.21; in Eggeling, *op. cit.*, vol. III, p. 133.
[3] *Taittirīya Saṃhitā* VI.1.6.4; in Keith, *Taittiriya Sanhita*, vol. II, p. 493.
[4] *Kātyāyana-Śrauta-Sūtra* XXII.2.9; in Ranade, *Kātyāyana Śrauta Sūtra*, p. 561; Keith, *Taittiriya Sanhita*, vol. I, p. cxv.
[5] *Baudhāyana-Śrauta-Sūtra* XVII.32; in Dandekar, *op. cit.*, vol. I, pt. 2, p. 904.

called) 'kuvala, karkandhu, or badara.' He (Indra) became emptied out of everything, for Soma is everything. Being thus purged by Soma, he walked about as one tottering. The Asvins cured him by this (offering), and caused him to be supplied with everything, for Soma is everything. By offering he indeed became better. The gods spake, 'Aha! these two have saved him, the well-saved (sutrâta):' hence the name Sautrāmaṇī.[1]

The *Śatapatha-Brāhmaṇa* continues:

Indra slew Tvashtri's son, Visvarûpa. Seeing his son slain, Tvashtri exorcized him (Indra), and brought Soma-juice suitable for witchery, and withheld from Indra. Indra by force drank off his Soma-juice, thereby committing a desecration of the sacrifice. He went asunder in every direction, and his energy, or vital power, flowed away from every limb. From his eyes his fiery spirit flowed, and became that grey (smoke-coloured) animal, the he-goat; and what (flowed) from his eyelashes became wheat, and what (flowed) from his tears became the kuvala-fruit. From his nostrils his vital power flowed, and became that animal, the ram; and what (flowed) from the phlegm became the Indra-grain, and what moisture there was that became the badara-fruit. From his mouth his strength flowed, it became that animal, the bull; and what foam there was became barley, and what moisture there was became the karkandhu-fruit...From his navel his life-breath flowed, and became lead,—not iron, nor silver; from his seed his form flowed, and became gold; from his generative organ his essence flowed, and became parisrut (raw fiery liquor); from his hips his fire flowed, and became surâ (matured liquor), the essence of food...From his hair his thought flowed, and became millet...from his marrow his drink, the Soma-juice, flowed, and became rice...Now at that time he (Indra) had to do with Namu*k*i, the Asura. Namu*k*i bethought him, 'He has been undone once for all: I will seize upon his energy, his vital power, his Soma-drink, his food.' By (taking) that Surâ-liquor of his he seized upon his energy, or vital power, his Soma-drink, his food. He lay there dissolved. The gods gathered around him, and said, 'Verily,

[1] *Śatapatha-Brāhmaṇa* V.5.4.2-12; in Eggeling, *op. cit.*, vol. III, pp. 130-131.

> he was the best of us; evil has befallen him: let us heal him!'...The two A*s*vins and Sarasvatî, having taken the energy, or vital power, from Namu*k*i, restored them to him (Indra), and saved him from evil. 'Truly, we have saved him from evil so as to be well-saved (sutrâta),' they thought, and this became the Sautrāmaṇī...[1]

The text continues:

> Verily, his fiery spirit, his energy, or vital power, depart from him whom Soma purges either upwards or downwards. As to this they say, 'Truly, the Soma-juice is the Brâhma*n*a's food; and, indeed, it is not owing to Soma when a Brâhma*n*a vomits Soma; and he who vomits Soma is one who, whilst being fit to (gain) prosperity, does not gain prosperity, and who, whilst being fit to (gain) cattle, does not gain cattle, for Soma is cattle.' Let him seize for sacrifice that grey (he-goat) of the A*s*vins, the ram of Sarasvatî, and the bull of Indra; for the A*s*vins are the physicians of the gods, and it is by them that he heals this (Sacrificer); and Sarasvatî is healing medicine, and it is with her help that he prepares medicine for him; and Indra is energy (indriya)...There are grains of rice and grains of millet, grains of wheat and kuvala jujubes, Indra-grain and badara jujubes, grains of barley and karkandhu jujubes, malted rice and barley: both cultivated and wild-grain food he thereby secures; and by means of both kinds of food he duly lays energy and vital power into his own self. With lead he buys the malted rice, with (sheep's) wool the malted barley, with thread the (fried) rice-grain,—that lead is a form of both iron and gold, and the *S*autrāmaṇī is both an ish*t*i-offering and an animal sacrifice...Here now, other Adhvaryus buy the malted rice with lead from a eunuch, saying 'That is that; for the eunuch is neither woman nor man, and the Sautrāmaṇī is neither an ishti-offering nor an animal sacrifice.' But let him not do so, for the Sautrāmaṇī is both an ishti and an animal sacrifice, and the eunuch is something unsuccessful among men...Let him rather buy them from a vendor of Soma, for the Sautrāmaṇī is Soma...[2]

The text continues:

[1] *Śatapatha-Brāhmaṇa* XII.7.1.1-14; in Ibid., vol. V, pp. 213-217.
[2] *Śatapatha-Brāhmaṇa* XII.7.2.1-12; in Ibid., vol. V, pp. 217-220.

> By means of the Surâ-liquor Namu*k*i, the Asura, carried off Indra's (source of) strength, the essence of his food, the Soma-drink… [Indra] struck off the head of Namu*k*i…In his (Namu*k*i's) severed head there was the Soma-juice mixed with blood. They loathed it. They perceived that (means of) drinking separately (one of) the two liquids,—'King Soma, the drink of immortality, is pressed;'—and having thereby made that (Soma) palatable, they took it in (as food). With, 'Thee, the sweet (liquor I mix) with the sweet (Soma),' he compounds (the ingredients for the preparation of) the Surâ-liquor, and makes it palatable; —'the strong with the strong,' he thereby bestows energy on him (the Sacrificer);—'the immortal with the immortal,' he thereby bestows life on him;—'the honeyed with the honeyed,' he thereby bestows flavour to it (the liquor);—'I mix with the Soma,' he thereby makes it (the Surâ-liquor) a form of Soma. 'Thou art Soma: get thee matured for the A*s*vins! get thee matured for Sarasvatî! get thee matured for Indra Sutrâman!' for these were the deities who first prepared that sacrifice, and with their help he now prepares it; and, moreover, he thereby provides these deities with their share. He distils it with a view to (its being like) the Soma-pressing. For three nights it remains standing, for the Soma remains standing for three nights after it has been bought: he thus makes it a form of Soma…With, 'Purified by Vâyu's purifier is the forward-flowing, exceeding swift Soma,' he purifies (the liquor)… 'with the Surâ the Soma,' for with the Surâ-liquor is Soma;—'the juice, is distilled,' for from the distilled the juice is obtained;—'for joy,' to joy (intoxication), indeed, the Soma-juice contributes, and to joy also does the Surâ-liquor: he thus secures both the joy of the Soma, and the joy of the Surâ…With 'Yea, even as the owners of barley cut their barley…,' (the Adhvarhu) fills (three) cups of milk,—barley-stalks are Soma-stems, and milk is Soma-juice: by means of Soma he thus makes it Soma-juice.[1]

The identification of barley with Soma is corroborated in the *Atharva-Veda*: "The grains of rice, of barley, that are scattered out—those are soma-

[1] *Śatapatha-Brāhmaṇa* XII.7.3.1-13; in Ibid., vol. V, pp. 222-227.

shoots."[1] The *Kauṣītaki-Brāhmaṇa* states: "Now we have called the victim Soma; and so the cakes. Ten are they, shoots of Soma; the old shoot, which they press here; the glad shoot, the waters; the sap shoot, rice; the male shoot, barley; the bright shoot, milk; the living shoot, the victim; the immortal shoot, gold; the Ṛc shoot; the Yajus shoot; the Sāman shoot; these are the ten Soma shoots; when all these unite, then is there Soma, then the pressed (Soma)."[2]

According to the *Śatapatha-Brāhmaṇa*, the gods and Asura demons both sprung from Pragâpati, and in their contests the gods were victorious by means of the barley that remained in their possession: "The gods then prevailed: by means of these (barley-grains)…They said, 'Come, let us put into the barley whatever sap there is of all plants!' And, accordingly, whatever sap there was of all plants, that they put into the barley."[3] The text states: "to Pragâpati belong these two (saps of) plants, to wit the Soma and the Surâ."[4] The *Kauṣītaki-Brāhmaṇa* states: "Soma is the king of the plants… he gives the honey drink, it is because this is the sap of the forest things."[5] The *Śatapatha-Brāhmaṇa* continues:

> Pragâpati created the (Soma-)sacrifice. He took it and performed it. When he had performed it, he felt like one emptied out. He saw this sacrificial performance, the Sautrâma*n*î, and performed it, and then he was again replenished; and, indeed, he who performs the Soma-sacrifice is, as it were, emptied out, for his wealth, his prosperity is, as it were taken from him. Having performed the Soma-sacrifice one ought to perform the Sautrâma*n*î…Suplan Sâr*n*gaya asked Pratîdar*s*a Aibhâvata,[6] 'Seeing that neither does one become initiated, nor are Soma-shoots thrown down (to be pressed), how then does the Sautrâma*n*î become a Soma-sacrifice?' He replied…'And, indeed, those (materials) are the Soma-shoots at this sacrifice,' they say, 'to wit, the malted rice, the malted barley, and the fried rice.' The malted rice, indeed, is of the form of the morning pressing…And the malted barley is of the form of the

[1] *Atharva-Veda* IX.6.14; in Whitney, *op. cit.*, vol. II, p. 540.
[2] *Kauṣītaki-Brāhmaṇa* XIII.4; in Keith, *Rigveda Brahmanas*, p. 420.
[3] *Śatapatha-Brāhmaṇa* III.6.1.8-10; in Eggeling, *op. cit.*, vol. II, p. 142.
[4] *Śatapatha-Brāhmaṇa* V.1.2.10; in Ibid., vol. III, p. 8.
[5] *Kauṣītaki-Brāhmaṇa* IV.12; in Keith, *Rigveda Brahmanas*, p. 369.
[6] The latter was the king of the *S*vikna, and the former his student. Eggeling, *op. cit.*, vol. V, p. 239 n. 2.

midday-pressing...And the fried rice is of the form of the evening-pressing...As to this they say, 'That Sautrâma*n*î, surely, is of the form of both effused (extracted) and infused (Soma);— to wit, that essence of both water and plants, the milk, is of the form of the effused (Soma); and that essence of food, the liquor, is of the form of infused (Soma): by both (kinds of) pressings he thus expresses it, by both pressings he secures it. As to this they say, 'Seeing that the Soma-juice is pressed out by stones, how as to the Sautrāmaṇī?' Let him reply, 'By the directions [praishas]'...All (the praishas) contain (the word) 'payas' (milk), for in the form of milk Soma is (here) pressed; they all contain (the word) 'Soma,' for the sake of (conformity with) the Soma-pressing; they all contain (the word) 'parisrut' (spirituous liquor), for in the form of spirituous liquor Soma is (here) pressed...they all contain (the word) 'madhu' (honey), for this—to wit, honey—is manifestly a form of Soma... This Sautrāmaṇī, then, is manifestly a Soma-sacrifice...[1]

The prescribed formulas for the performance of the Sautrāmaṇī include the following from the *Vājasaneyi-Saṃhitā*:

Sweet with the sweet, I sprinkle thee with Soma, strong with the strong, the nectar with the nectar,

The honey-sweet with what is sweet as honey.

Soma art thou...

Soma with Wine, pressed, filtered for the banquet...

What then ? As men whose fields are full of barley reap the ripe corn...

For each of you is made a God-appointed place...

Surâ the strong art thou. This here is Soma...

Symbols of Dîkshâ are grass buds, of Prâyaṇîya sprouts of corn,[2]

Of Soma-purchasing fried grains are symbols, Soma-shoots and meath (honey).

[1] *Śatapatha-Brāhmaṇa* XII.8.2.1-21; in Eggeling, *op. cit.*, vol. V, pp. 239-245.

[2] Dîkshâ is reference to consecration in the Soma-sacrifice, and Prâyaṇîya is an introductory libation in the Soma rite. Ralph T. H. Griffith (trans.), *The Texts of the White Yajurveda*, Benares, E. J. Lazarus and Co., 1899, pp. 35 n. 6, 174 n. 13.

Âtithya's sign is Mâsara, the Gharma's symbol Nagnahu.[1]

Three nights with Surâ poured, this is the symbol of the Upasads.[2]

Emblem of purchased Soma is Parisrut, foaming drink effused: Indra's balm milked for Indra by the Aṣvins and Sarasvatî…

Grain roasted, gruel, barley-meal, grains of rice roasted, milk and curd

Are types of Soma: mingled milk, sweet whey, of sacrificial food.

Type of parched corn is jujube-fruit; wheat of the roasted grains of rice; Jujube the type of barley-meal, and Indra-grains of gruel-groats.

Symbol of milk are barley-grains, symbol of curd are jujube-fruits.

Whey is the type of Soma, the milk-mixture type of Soma's pap…[3]

The rite with sacred grass, wine, store of heroes, the mighty ones speed on with adorations.

May we, sweet-singing sacrificers, setting Soma mid Gods in heaven, give joy to Indra.

All essence of thine own in plants collected, all strength of Soma when poured out with Surâ—

Therewith impel with joy the sacrifice, Sarasvatî, the Aṣvíns, Indra, Agni.

That which Sarasvatî poured out for Indra, by Aṣvins brought from Namuchi the demon,

This flowing drop, brilliant and full of sweetness, I drink and feed on here, the King, the Soma.[4]

Further, the *Vājasaneyi-Saṃhitā* says:

---

[1] Âtithya is the ceremonial reception of Soma when brought to the place of sacrifice. Gharma is the cauldron for boiling libations in the Soma-sacrifice, and Nagnahu is barley flour. Ibid., p. 174 n. 14.

[2] The Upasads are offerings of butter to Agni, Soma, and Vishṇu thrice daily in the Soma-sacrifice. Ibid., p. 35 n. 8.

[3] Pap, or *caru*, is the oblation of boiled rice or barley mixed with butter and milk in the Soma-sacrifice. Ibid., p. 175 n. 23.

[4] *Vājasaneyi-Saṃhitā* XIX.1-34; in Griffith, *White Yajurveda*, pp. 172-177.

Indra, with waters' foam didst thou wrench off the head of Namuchi,

Subduing all contending hosts.

King Soma, pressed, the Drink of Life, left Death behind with Soma-dregs.

By Law came truth and Indra-power, the pure bright drinking-off of juice. The power of Indra was this sweet immortal milk...

Prajâpati by Brahma drank the essence from the foaming food, the princely power, milk, Soma juice...

By holy lore Prajâpati drank up both forms, pressed and unpressed...

Seeing the foaming liquor's sap, Prajâpati with the bright drank out the bright, the milk, the Soma juice...

This his immortal shape with mighty powers three Deities bestowing gifts compounded.

His hair they made with sprouts of grass and barley, and roasted grain with skin and flesh supplied him.

His inner shape Sarasvatî arranges and, borne on bright paths, the Physician Aṣvins:

With Mâsaras and sieve his bone and marrow, as on the Oxen's hide they lay the liquor.

By thought Sarasvatî with both Nâsatyas (Aṣvins) forms lovely treasure and a beauteous body.

Like shuttle through the loom the steady ferment mixes the red juice with the foaming spirit...

With wheat eyelashes and with jujube eyebrows they clothe as 'twere a black and brilliant figure...

Barley and sacred grass composed his eyebrows: from his mouth came the jujube and sweet honey...

Splendour of victims, powerful oblation, honey and meath with milk and foaming liquor,

Healing Sarasvatî effused, and Aṣvins; from pressed and unpressed Soma, deathless Indu.[1]

The text continues:

[1] *Vājasaneyi-Saṃhitā* XIX.71-95; in Ibid., pp. 182-185.

Fain would I know that holy world where want and languor are unknown,

Where in complete accordance move Indra and Vâyu side by side.

Let thy shoot be united with his tendril, joint combine with joint.

Imperishable sap for joy, thine odour be the Soma's guard !

They pour it out, they sprinkle it, they scatter it, they make it pure.

In the brown Surâ's ecstasy he says What art thou ?...

Adhvaryu, on the straining-cloth pour thou the Soma pressed with stones:

Purify it for Indra's drink...

The Aṣvins brought from Namuchi pressed Soma bright with foaming juice.

Sarasvatî with sacred grass brought that to Indra for his drink...

As drink to gladden Indra, poured strong Soma with the foaming juice.

The Aṣvins, our Sarasvatî, and Tvashṭar, when the juice was shed,

Gave Indra balm, yea, mead as balm, glory and fame and many a shape.

Praising the foaming liquor at due times, Indra, Vanaspati, Sarasvatî as cow gave for sweet beverage with the Aṣvins Twain.

Aṣvins, to Indra ye with cows, with Mâsara and foaming drink

Gave, with Sarasvatî—All hail!—the pressed out Soma juice and mead.[1]

The *Vājasaneyi-Saṃhitā* goes on:

Let the Hotar sacrifice with fuel to Agni in the place of libation, to the Aṣvins, Indra, Sarasvatî. A grey-coloured he-goat with wheat, jujube-fruit and sprouts of rice becomes a sweet salutary remedy, splendour, might, milk, Soma. Let them enjoy sweet butter with foaming liquor...

Let the Hotar, Tanûnapât, worship Sarasvatî. A sheep, a ram,

[1] *Vājasaneyi-Saṃhitā* XX.26-66; in Ibid., pp. 189–193.

a salutary remedy on the honey-sweet path, bearing to the Aṣvins and Indra heroic strength, with jujube-fruit, Indra-grains, sprouts of rice, becomes a salutary remedy, milk, Soma. Let them enjoy…

Let the Hotar worship Narâsamsa and the Lord Nagnahu. A ram with Surâ a salutary remedy, Sarasvatî the Physician, the golden car of the Aṣvins, the victim's omentum, with jujube-fruit, Indra-grains, and rice-sprouts, become a salutary remedy, the manly strength of Indra, milk, Soma. Let them enjoy…

Let the Hotar, magnified with oblations, offering sacrifice, worship Sarasvatî and Indra, increasing them with strength, with a bull and a cow. Strength and medicine to the Aṣvins and Indra are meath with jujube-fruit, Mâsara with parched grain, milk, Soma. Let them enjoy…

Let the Hotar worship Tvashṭar full of good seed, the Bull active for men, Indra, the Aṣvins, Sarasvatî the Physician. Vigour, speed, power, a fierce wolf as physician, fame with Surâ is a medicine, Mâsara with grace. Milk, Soma. Let them enjoy…

Let the Hotar worship the Aṣvins, Sarasvatî, and Indra the Good Deliverer. These your Somas, pressed, rejoicing with goats, rams, bulls, giving pleasure with rice-shoots, young blades of corn, parched grain, joy-givers adorned with Mâsara, bright, milky, immortal, presented, dropping honey…Let them drink, rejoice in, enjoy the Soma meath.[1]

The Aṣvins are praised in the *Ṛgveda* for drinking Surāma, that is, Surā mixed with Soma, and helping Indra in his battle with Namuci the Asura.[2] Surāma was also an illness caused by drinking Surā in excess, from which Indra suffered in the Namuci legend, and this term later became an epithet of Soma.[3] In the *Sāma-Veda*, Indra is once more said to drink Soma mixed with fermented barley: "Indra whose jaws are strong hath drunk of worshipping

---

[1] *Vājasaneyi-Saṃhitā* XXI.29-42; in Ibid., pp. 199-201.

[2] *Ṛgveda* X.131.4-5; in Griffith, *Rgveda*, vol. II, p. 578; Van Nooten and Holland, *op. cit.*, p. 555; Hillebrandt, *op. cit.*, vol. I, pp. 321, 324; Pandurang Vaman Kane, *History of Dharmaśāstra (Ancient and Mediæval Religious and Civil Law)*, Second Edition, Poona, Bhandarkar Oriental Research Institute, 1974, vol. II, pt. 2, p. 792.

[3] Macdonell and Keith, *op. cit.*, vol. II, p. 459.

Sudaksha's draught, The Soma juice with barley brew."[1] The formulas for the Surā-sacrifice of the Caraka Sautrāmaṇī, performed in the Rājasūya Soma-sacrifice, include the following from the *Vājasaneyi-Saṃhitā*:

> Soma the Wind, purified by the strainer, Indra's meet friend, hath gone o'erflowing backward.
>
> What then ? As men whose fields are full of barley reap the ripe corn, removing it in order,
>
> So bring the food of these men, bring it hither, who pay the Sacred Grass their spoken Homage.
>
> Taken upon a base art thou. Thee for the Aṣvins.
>
> Thee for Sarasvatî, and thee for Indra, for the Excellent Protector.
>
> Ye Aṣvins, Lords of Splendour, drank full draughts of grateful Soma juice,
>
> And aided Indra in his deeds with Namuchi of Asura birth.[2]

The *Śāṅkhāyana-Śrauta-Sūtra* prescribes the following formula to accompany the drinking of Surā in the Rājasūya: "The mantra to accompany the partaking of the surā is: 'The (Soma), which the Aśvins took from the asuric Namuci, which Sarasvatī pressed out to obtain strength, this king Soma, the bright, sweet drop, I here partake of.'"[3] In the instructions contained in the *Aitareya-Brāhmaṇa* for the performance of the Rājasūya, Surā is called "a symbol of the lordly power,"[4] and the priest recites the following mantra while handing a bowl of Surā to the sacrificer:

> "With thy sweetest, most intoxicating
> Stream be thou purified, O Soma,
> Pressed for Indra to drink."

He should drink it (saying)

> "That which is left over of the pressed juice, rich in sap
> Which Indra drank mightily
> Here with auspicious mind this of him,
> I partake of Soma, the King.

---

[1] *Sāma-Veda* I.ii.ii.1.1; in Ralph T. H. Griffith (trans.), *The Hymns of the Sāmaveda*, Fourth Edition, Varanasi, Chowkhamba Sanskrit Series Office, 1963, p. 31.

[2] *Vājasaneyi-Saṃhitā* X.31-33; in Griffith, *White Yajurveda*, pp. 85-86.

[3] *Śāṅkhāyana-Śrauta-Sūtra* XV.15.13; in Caland, *Śāṅkhāyana-Śrautasūtra*, p. 437.

[4] *Aitareya-Brāhmaṇa* VIII.8; in Keith, *Rigveda Brahmanas*, p. 325.

To thee, O bull (the Soma) being pressed,
I offer the pressed juice to drink;
Rejoice and make thyself glad."

The Soma drink which is in the Surā is what is drunk by the Kṣatriya when anointed by this great anointing of Indra; not the Surā. Having drunk it he should address it with "We have drunk the Soma"[5] and "Be thou propitious to us."[6]

[5] "We have drunk Soma, we have become immortal, we have entered into light, we have known the gods." *Ṛgveda* VIII.48.3; quoted in Macdonell, *Vedic Mythology*, p. 109.

[6] *Aitareya-Brāhmaṇa* VIII.20; in Keith, *Rigveda Brahmanas*, p. 335.

Chapter 6: Mead and Honeydew of Ergot

In the Tantric rituals, Surā and Madya (wine) are regarded as Amrita and the elixir of immortality.[1] The Vedic term Amrita etymologically corresponds to Greek *ἀμβροσία* (ambrosia),[2] and is frequently used for Soma.[3] The Vedic term Madhu also denotes Soma.[4] Madhu is variously translated as honey or mead in the Vedic texts, and can also denote wine in the Indo-European languages.[5] The Vedic term Madhu is etymologically identical to Old English *medu*, "mead," and Greek *μέθυ* (*méthu*), "wine."[6] In the *Ṛgveda*, the Soma brought to Indra by an eagle is called mead (*mádhu*),[7] and Soma is also called the mead (*mádhu*) of Tvashṭar.[8] The priests who composed the *Ṛgveda* also

[1] Mitra, *op. cit.*, pp. 14-17; N. N. Bhattacharyya, *History of the Tantric Religion: A Historical, Ritualistic and Philosophical Study*, New Delhi, Manohar, 1982, pp. 141-145.

[2] Müller, *Biographies of Words*, p. 189.

[3] *Ṛgveda* V.2.3, VI.37.3, IX.62.6, IX.70.4, IX.74.4, X.186.3; in Griffith, *Rgveda*, vol. I, pp. 467, 597, vol. II, pp. 309, 327, 332, 607.

[4] *Ṛgveda* I.19.9, II.36.4, III.43.3, IV.18.13; in Ibid., vol. I, pp. 23, 306, 364, 418; Van Nooten and Holland, *op. cit.*, pp. 11, 143, 159, 181.

[5] Macdonell and Keith, *op. cit.*, vol. II, pp. 123-124; Calvert Watkins (ed.), *The American Heritage Dictionary of Indo-European Roots*, Second Edition, Boston, Houghton Mifflin Company, 2000, p. 52; J. P. Mallory and D. Q. Adams, *The Oxford Introduction to Proto-Indo-European and the Proto-Indo-European World*, Oxford, Oxford University Press, 2006, pp. 260-265.

[6] Henry George Liddell and Robert Scott, *A Greek-English Lexicon*, Ninth Edition, Oxford, Clarendon Press, 1996, p. 1091.

[7] *Ṛgveda* III.43.7, IV.18.13; in Griffith, *Rgveda*, vol. I, pp. 365, 418; Van Nooten and Holland, *op. cit.*, p. 181.

[8] *Ṛgveda* I.117.22; in Griffith, *Rgveda*, vol. I, p. 160; Van Nooten and Holland, *op. cit.*, p. 71; Macdonell, *Vedic Mythology*, p. 52.

employ the formula *somyám mádhu*, "Soma mead."[1] According to the text, the Maruts are "fond of honey,"[2] and drink "the delightful mead,"[3] which is Soma: "O ye whose gifts are cheering, coming to drink the (juice of the Soma) flowers...O Maruts, who never miss the Soma mead, hail to you here to enjoy yourselves."[4] The *Ṛgveda* also states: "Adhvaryus, make the sweet libations ready...we give thee of the mead to make thee joyful...The stalk hath poured, fair with spreading branches, the mead's bright glittering juice that dwells on mountains. The Soma hath been pressed for thee."[5] The text further connects Soma and mead: "I have partaken wisely of the sweet food... The food round which all deities and mortals, Calling it honey-mead (*mádhu*), collect together...We have drunk Soma and become immortal."[6] The *Sāma-Veda* states: "The living drops of Soma juice pour, as they flow...dropping meath. Soma, while thou art cleansed, most dear and watchful in the sheep's long wool...thou hast become a sage. Sprinkle our sacrifice with mead !"[7]

There are a few verses in the *Ṛgveda* that are thought to possibly indicate the mixing of honey with Soma, and even assuming that this doubtful interpretation is accurate, the use of honey was extremely rare in the Śrauta ritual; there is little evidence, if any, in the later literature for the mixing of honey with Soma.[8] The formulas frequently applied to Soma: *sómo mádhumān, somyám mádhu, mádiram mádhu/sómam* (*madirá-*, "intoxicating"), have been connected by Calvert Watkins to honeydew of ergot and barley.[9] The *Atharva-Veda* declares of the sacred barley sown on the legendary Sarasvatī River: "This barley, combined with honey, the gods plowed much on the Sarasvatī."[10] The *Ṛgveda* calls "the presser (of Soma), the Madhu-presser,"[11] and the *Aitareya-Brāhmaṇa* states: "O Adhvaryu, shalt thou press for Indra

[1] Macdonell, *Vedic Mythology*, p. 105.
[2] *Ṛgveda* VII.57.1; in Müller, *Vedic Hymns*, vol. I, p. 379.
[3] *Ṛgveda* V.61.11; in Ibid., vol. I, p. 356.
[4] *Ṛgveda* VII.59.5-6; in Ibid., vol. I, p. 386.
[5] *Ṛgveda* V.43.3-5; in Griffith, *Rgveda*, vol. I, p. 508.
[6] *Ṛgveda* VIII.48.1-3; in Macdonell, *Rigveda*, pp. 79-80; Van Nooten and Holland, *op. cit.*, p. 383.
[7] *Sāma-Veda* I.vi.i.3.8-9; in Griffith, *Sāmaveda*, p. 104.
[8] Hillebrandt, *op. cit.*, vol. I, pp. 318-320.
[9] Calvert Watkins, "Let Us Now Praise Famous Grains," *Proceedings of the American Philosophical Society*, 122 (1978), p. 13.
[10] *Atharva-Veda* VI.30.1; in Whitney, *op. cit.*, vol. I, p. 302.
[11] *Ṛgveda* IV.3.3; in Hermann Oldenberg (trans.), *Vedic Hymns*, Oxford, Clarendon Press, 1897, vol. II, p. 325.

the Soma rich in honey."[1] In the *Ṛgveda*, the Soma juice is also described as "honied" when pressed.[2] The pressed Soma juice is further described in the text as a wave of honey,[3] and the pressed juice flows with a stream of honey like a rain-charged cloud.[4]

Indra consumes three lakes of Soma in the *Ṛgveda*.[5] In two hymns addressed to the Maruts, the Storm-gods of the *Ṛgveda*, the text states: "O Maruts…the casks (clouds) on your chariots trickle everywhere, and you pour out the honey-like fatness (the rain) for him who praises you;"[6] and, "O Maruts…(the clouds) yielded three lakes (from their udders) as mead (*mádhu*) for the wielder of the thunderbolt (Indra)."[7] In the *Atharva-Veda*, honey and Surā are said to fill the mountains and streams: "With honey I mix the streams; the rugged mountains (are) honey;"[8] and, "let all these streams come unto thee, swelling honeyedly in the heavenly world…Having pools of ghee (butter), having slopes of honey, having strong drink (*súrā*) for water, filled with milk, with water, with curds."[9]

Surā and honey are also closely associated with the Aṣvins in the same text: "What glory [is] in the mountain…in gold, in kine, in strong-drink when poured out, [what] honey in sweet-drink, [be] that in me. O ye Aṣvins, lords of beauty ! anoint me with the honey of bees, that I may speak brilliant words among the people."[10] In two identical verses from consecutive hymns in the *Ṛgveda*, the Aṣvins pour out a hundred jars of Surā, and a hundred jars of honey: "Ye poured forth from the hoof of your strong charger a hundred jars of wine (*súrā-*) as from a strainer;"[11] and, "When from the hoof of your

[1] *Aitareya-Brāhmaṇa* II.20; in Keith, *Rigveda Brahmanas*, p. 149.

[2] *Ṛgveda* IV.45.5, IX.97.14; in Griffith, *Rgveda*, vol. I, p. 451, vol. II, p. 359; Van Nooten and Holland, *op. cit.*, pp. 195, 466.

[3] *Ṛgveda* III.47.1, IX.85.10, IX.86.2; in Griffith, *Rgveda*, vol. I, p. 367, vol. II, p. 341; Van Nooten and Holland, *op. cit.*, pp. 161, 459.

[4] *Ṛgveda* IX.2.9; in Griffith, *Rgveda*, vol. II, p. 270; Van Nooten and Holland, *op. cit.*, p. 421.

[5] *Ṛgveda* V.29.7-8; in Griffith, *Rgveda*, vol. I, p. 489.

[6] *Ṛgveda* I.87.2; in Müller, *Vedic Hymns*, vol. I, p. 159.

[7] *Ṛgveda* VIII.7.9-10; in Ibid., vol. I, p. 390; Van Nooten and Holland, *op. cit.*, p. 342.

[8] *Atharva-Veda* VI.12.3; in Whitney, *op. cit.*, vol. I, p. 289.

[9] *Atharva-Veda* IV.34.5-6; in Ibid., vol. I, pp. 206-207.

[10] *Atharva-Veda* VI.69.1-2; in Ibid., vol. I, p. 333.

[11] *Ṛgveda* I.116.7; in Griffith, *Rgveda*, vol. I, p. 155; Van Nooten and Holland, *op. cit.*, p. 69.

strong horse ye showered a hundred jars of honey (*mádhū-*) for the people."[1] The Soma formula is completed in the same hymn: "Ploughing and sowing barley, O ye Aṣvins, milking food for men…(Dadhyach)[2] revealed to you… sweet Soma (*mádhu*, mead), Tvashṭar's secret."[3] And elsewhere in the text: "Ye (Aṣvins) with your plough…ploughed the first harvest in the sky…This Soma pressed with stones is yours, ye Heroes."[4]

The Aṣvins also possess a honey-whip (*madhukaśā́*), which the Vedic literature connects with Soma, and Watkins connects with honeydew of ergot and barley.[5] In a hymn dedicated to the Aṣvins' honey-whip, the *Atharva-Veda* formulaically connects honeydew and barley with Surā and Soma:

> From heaven, from earth, from the atmosphere, from the sea, from the fire, and from the wind, the honey-lash hath verily sprung. This, clothed in am*ri*ta (ambrosia), all the creatures revering, acclaim in their hearts.
>
> Great sap of all forms (colours) it hath—they call thee moreover the seed of the ocean. Where the honey-lash comes bestowing gifts, there life's breath, and there immortality has settled down…
>
> Mother of the Âdityas, daughter of the Vasus, breath of life of created beings, nave of immortality, the honey-lash, golden-coloured, dripping ghee, as a great embryo, moves among mortals.
>
> The god's begot the lash of honey, from it came an embryo having all forms (colours)…
>
> Who knows it and who perceives it, the inexhaustible, soma-holding cup that has come from the heart of it (the honey-lash) ? 'Tis the wise priest: he shall derive inspiration from it !…
>
> The thunder is thy voice, O Pragâpati; as a bull thou hurlest thy fire upon the earth. From the fire, and from the wind the

1 *Ṛgveda* I.117.6; in Griffith, *Rgveda*, vol. I, p. 158; Van Nooten and Holland, *op. cit.*, p. 70.

2 Dadhyach was the son of Atharvan, the priest who first obtained fire and offered prayer and Soma libations to the gods. The Atharvans, or priests of fire and Soma, performed the duties of the Adhvaryus. Griffith, *Sāmaveda*, pp. 3 n. 9, 119 n. 2, 170 n. 1.

3 *Ṛgveda* I.117.21-22; in Griffith, *Rgveda*, vol. I, p. 160; Van Nooten and Holland, *op. cit.*, p. 71.

4 *Ṛgveda* VIII.22.6-8; in Griffith, *Rgveda*, vol. II, p. 153.

5 Watkins, "Praise Famous Grains," p. 13.

honey-lash hath verily sprung, the strong child of the Maruts.

As the soma at the morning-pressure is dear to the Aṣvins, thus in my own person, O Aṣvins, lustre shall be sustained !...

As bees carry together honey upon honey, thus in my own person, O Aṣvins, lustre shall be sustained !

As the bees pile this honey upon honey, thus in my own person, O Aṣvins, lustre, brilliance, strength, and force shall be sustained !

The honey that is in the mountains, in the heights; in the cows, and in the horses; the honey which is in the surâ (brandy) as it is being poured out, that shall be in me![6]

O Aṣvins, lords of brightness, anoint me with the honey of the bee, that I may speak forceful speech among men !...

He that knoweth the seven honies of the whip becomes rich in honey; (to wit), the Brâhma*n*a, the king, the cow, the ox, rice, barley, and honey as the seventh.[7]

The *Ṛgveda* states: "Nigh to us come the Aṣvins' lauded three-wheeled car, the car laden with meath and drawn by fleet-foot steeds...bestowing all delight...Bring hither nourishment for us, ye Aṣvins Twain; sprinkle us with your whip that drops with honey-dew."[8] And elsewhere in the text:

Waken the Aṣvin Pair who yoke their car at early morn: may they
Approach to drink this Soma juice.
We call the Aṣvins Twain, the Gods borne in a noble car, the best
Of charioteers, who reach the heavens.
Dropping with honey is your whip, Aṣvins, and full of pleasantness:
Sprinkle therewith the sacrifice.
As ye go thither in your car, not far, O Aṣvins, is the home
Of him who offers Soma juice.[9]

Cultivated cereals, particularly barley, held a central role in early Vedic

[6] Bloomfield's identification here of Surā with brandy, an alcoholic beverage distilled from wine or fermented fruit juice, is technically inaccurate.
[7] *Atharva-Veda* IX.1.1-22; in Bloomfield, *op. cit.*, pp. 229-232.
[8] *Ṛgveda* I.157.3-4; in Griffith, *Rgveda*, vol. I, p. 209.
[9] *Ṛgveda* I.22.1-4; in Ibid., vol. I, p. 25.

religion, and this statement also holds true in regard to ancient Greece. Watkins has convincingly demonstrated that the formula used by the Vedic priests to produce the Soma beverage was the same formula employed by the priesthood in Greece that performed the Eleusinian rites.[1] The *Homeric Hymn to Demeter* records the ingredients of the mixture consumed by the initiates of the Greater Mystery at Eleusis, which only contained "barley and water…[with] *glechon*."[2] In Homer's *Odyssey*, Circe's potion also contains barley and drugs, mixed in wine: "[She] made for them a potion of cheese and barley meal and yellow honey with Pramnian wine; but in the food she mixed baneful drugs."[3] In Homer's *Iliad*, the goddess Athena is offered libations of "honey-sweet wine,"[4] and the potion mixed by Hecamede also contains barley and honey in wine: "Hecamede mixed a potion…pale honey, and ground meal of sacred barley; and beside them a beauteous cup…Therein the woman, like to the goddesses, mixed a potion for them with Pramnian wine, and on this she grated cheese of goat's milk with a brazen grater, and sprinkled thereover white barley meal; and she bade them drink, when she had made ready the potion."[5]

In Homer's *Iliad*, Hector addresses his horses, fed by Andromache, who "set before you honey-hearted wheat, and mingled wine for you to drink,"[6] and the horses of Diomedes are also fed "honey-sweet corn."[7] In Ovid's *Metamorphoses*, the horses harnessed to the chariot of the Sun are fed ambrosia: "Beneath the western skies lie the pastures of the Sun's horses. Here not common grass, but ambrosia is their food."[8] A similar statement is found in the *Vājasaneyi-Saṃhitā*: "United with the Soma, ye, for Indra, are corn for his two tawny steeds to feed on."[9] In a hymn addressed to Indra, whose flying-horses are fed following the Soma-pressing, the *Ṛgveda* states:

> Mount the Bay Horses to thy chariot harnessed…
> Thou, hastening to us, shalt drink the Soma…

[1] Watkins, "Praise Famous Grains," pp. 9-17.
[2] *Homeric Hymn to Demeter* 209-210.
[3] Homer, *Odyssey* X.233-236.
[4] Homer, *Iliad* X.579.
[5] Ibid. XI.624-641.
[6] Ibid.VIII.188-189.
[7] Ibid. X.569.
[8] Ovid, *Metamorphoses* IV.214-215.
[9] *Vājasaneyi-Saṃhitā* VIII.11; in Griffith, *White Yajurveda*, p. 61.

Bring the strong Steeds who drink the warm libation…

Let thy Steeds eat; set free thy Tawny Horses, and roasted grain like this consume thou daily…

The grass is strewn for thee, pressed is the Soma; the grain is ready for thy Bays to feed on.[1]

Composed in the seventh century A.D., the *Harṣacarita* mentions "holy Soma, the lord of the plants," the "elixir of life," the "land of the philosopher's stone," the "wood-nymphs dispensing ambrosia," the "ambrosia dew," the "cups bristling with dew-besprent blades of barley," the "mingled scent of wine, ambrosia," the white "ambrosia foam," and "cooking the ambrosial posset," an oblation of fermented barley or rice, milk, and ghee offered to the gods.[2] In the *Atharva-Veda,* Soma is once more identified with grain in the instructions for the preparation of the *brahmaudana,* the porridge made solely of boiled cereal that is given as a fee to the Brahman priests:

O Agni, come into being ! Aditi here in her throes, longing for sons, is cooking the porridge for the Brahmans. The seven *Ri*shis, that did create the beings, shall here churn thee, along with progeny !…

This great goddess earth, kindly disposed, shall receive the (sacrificial) skin ! Then may we go to the world of well-doing (heaven) !

Lay these two press-stones, well coupled, upon the skin; crush skillfully the (soma-) shoots for the sacrificer ! Crush down, (O earth), and beat down, those who are hostile to her (the wife); lift up high, and elevate her offspring !

Take into thy hands, O man, the press-stones that work together: the gods that are to be revered have come to thy sacrifice! Whatever three wishes thou dost choose, I shall here procure for thee unto fulfillment.

This, (O winnowing-basket), is thy purpose, and this thy nature: may Aditi, mother of heroes, take hold of thee ! Winnow out those who are hostile to this (woman); afford her wealth and

[1] *Ṛgveda* III.35.1-7; in Griffith, *Rgveda*, vol. I, p. 356.

[2] *Harṣacarita* 29, 82, 105, 107, 115, 163, 170, 223, 281; in E. B. Cowell and F. W. Thomas (trans.), *The Harṣa-carita of Bāṇa*, Delhi, Motilal Banarsidass, 1961, pp. 20, 61, 80, 82, 130, 137, 193, 251, 265.

undiminished heroes !

Do ye, (O grains), remain in the (winnowing-) basket, while (the wind) blows over you; be separated, ye who are fit for the sacrifice, from the chaff !...

Purified by (our) prayer, and clarified by the ghee are the soma-shoots, (and) these sacrificial grains. Enter the water; may the pot receive you ! When ye have cooked this (porridge) go ye to the world of the pious (heaven) !...

May this sacrificial ladle (sru*k*), the second hand of Aditi, which the seven *Ri*shis, the creators of the beings, did fashion, may this spoon, knowing the limbs of the porridge, heap it upon the altar !

The divine (Brâhma*n*as) shall sit down to thee, the cooked sacrifice: do thou again descending from the fire, approach them! Clarified by soma settle in the belly of the Brâhma*n*as; the descendants of the *Ri*shis who eat thee shall not take harm !

O king Soma, infuse harmony into the good Brâhma*n*as who shall sit about thee ! Eagerly do I invite to the porridge the *Ri*shis, descended from *Ri*shis, that are born of religious fervour, and gladly obey the call.

These pure and clear sacrificial women (the waters) I put into the hands of the Brâhma*n*as severally. With whatever wish I pour this upon you, may Indra accompanied by the Maruts grant this to me !

This gold is my immortal light, this ripe fruit of the field is my wish-granting cow. This treasure I present to the Brâhma*n*as: I prepare for myself a road that leads to the Fathers in the heavens.

Scatter the spelt into Agni *G*âtavedas (the fire), sweep away to a far distance the chaff! This (chaff) we have heard, is the share of the ruler of the house (Agni), and we know, too, what belongs to Nir*ri*ti (destruction) as her share.

Note, (O porridge), him that takes pains, and cooks and presses the soma; lift him up to the heavenly road, upon which, after he has reached the fullest age, he shall ascend to the highest firmament, the supreme heavens !

> Anoint (with ghee), O adhvaryu (priest), the surface of this sustaining (porridge), make skillfully a place for the melted butter; with ghee do thou anoint all its limbs ! I prepare for myself a road that leads to the Fathers in the heavens.[1]

The text also refers to the porridge as Amrita, the food of the gods:

> Around the water united, sit ye down, O children; around this living (father) and the waters that refresh the living ! Partake of these (waters), and of that porridge which the mother of you two cooks, and which is called am*r*ita (ambrosia) !
>
> The porridge which the father of you two, and which the mother cooks, unto freedom from defilement and foulness of speech, that porridge with a hundred streams (of ghee), leading to heaven, has penetrated with might both the hemispheres of the world.
>
> In that one of the two hemispheres and the two heavenly worlds, conquered by the pious, which especially abounds in light, and is rich in honey…
>
> All-embracing, about to be covered with ghee, enter, (O pot), as a co-dweller this space! Take hold of the winnowing-basket, that has been grown by the rain: the spelt and the chaff it shall sift out !
>
> Three regions are constructed after the pattern of the Brâhma*n*a: yonder heaven, the earth, and the atmosphere.—Take the (soma-) shoots, and hold one another, (O man and wife) ! They (the shoots) shall swell (with moisture), and again go back into the winnowing-basket !
>
> Of manifold variegated colours are the animals, one colour hast thou, (O porridge), when successfully prepared.—Push these (soma-) shoots upon this red skin; the press-stone shall purify them as the washer-man his clothes !…
>
> Whether ye are over-abundant or just sufficient, ye are surely clear, pure, and immortal: cook, ye waters, instructed by the husband and wife, obliging and helpful, the porridge !
>
> Counted drops penetrate into the earth, commensurate with

[1] *Atharva-Veda* XI.1.1-31; in Bloomfield, *op. cit.*, pp. 179-184.

the breaths of life and the plants. The uncounted golden (drops), that are poured into (the porridge), have, (themselves) pure, established complete purity.

The boiling waters rise and sputter, cast up foam and many bubbles. Unite, ye waters, with this grain, as a woman who beholds her husband in the proper season !

Stir up (the grains) as they settle at the bottom: let them mingle their inmost parts with the waters ! The water here I have measured with cups; measured was the grain, so as to be according to these regulations.

Hand over the sickle, with haste bring promptly (the grass for the barhis); without giving pain let them cut the plants at the joints ! They whose kingdom Soma rules, the plants, shall not harbour anger against us !

Strew a new barhis for the porridge: pleasing to its heart, and lovely to its sight it shall be ! Upon it the gods together with the goddesses shall enter; settle down to this (porridge) in proper order, and eat it !

O (instrument of) wood,[1] settle down upon the strewn barhis, in keeping with the divinities and the agnish*t*oma rites ![2] Well shaped, as if by a carpenter (Tvashtar) with his axe, is thy form. Longing for this (porridge) the (gods) shall be seen about the vessel !

In sixty autumns the treasurer (of the porridge) shall fetch it, by the cooked grain he shall obtain heaven; the parents and the children shall live upon it. Bring thou this (man) to heaven, into the presence of Agni !

(Thyself) a holder, (O pot), hold on to the foundation of the earth: thee, that art immoveable the gods (alone) shall move ! Man and wife, alive, with living children, shall remove thee from the hearth of the fire !

Thou hast conquered and reached all worlds; as many as are our wishes, thou hast satisfied them. Dip ye in, stirring stick and

[1] This is reference to a wooden platter.

[2] The agnish*t*oma is a Soma-sacrifice in honor of the Fire-god Agni.

spoon ! Place it (the porridge) upon a single dish !

Lay (ghee) upon it, let it spread forth, anoint this dish with ghee !...

The goodly streams, swelling with honey, mixed with ghee, the seats of ambrosia,[1] all these does he obtain, ascends to heaven.[2]

[1] Or "the navels of immortality." Bloomfield, *op. cit.*, p. 653; Whitney, *op. cit.*, vol. II, p. 690.

[2] *Atharva-Veda* XII.3.4-41; in Bloomfield, *op. cit.*, pp. 185-190.

Addendum

In his *Soma: Divine Mushroom of Immortality*, R. Gordon Wasson writes: "In a world where farming was already well developed one would expect Soma to be cultivated. A plant with properties so extraordinary would elicit the utmost attention, though owing to its sacred character we would rather expect its growth to be confined to the gardens of the higher priesthood. But the fact is that there is never a mention of its cultivation." To accommodate his candidate *A. muscaria*, Wasson concluded that the Soma-plant was not cultivated and only grew in the mountains: "The poets say that Soma grows high in the mountains...They never speak of it as growing elsewhere."[1] John Brough writes: "It seems thus to have been generally accepted that the Soma-plant did grow on mountains. But the facile deduction that it grew exclusively on mountains does not follow logically, and may be contradicted by two passages: *Yasna* 10.17...'I praise all the Haomas whether those on the heights of the mountains, whether those in the valleys of the rivers'; RV 8.6.28, where the *vipra* (Soma) is said to have been born...'in the hidden place [or on the slope?] of the mountains, and in the confluence of the rivers'. We may note also the name of the river Aṃśumatī, RV 8.96.13-15, the river 'abounding in Soma-plants'. This river appears only in a legendary context; but

[1] R. Gordon Wasson, *Soma: Divine Mushroom of Immortality*, New York, Harcourt Brace Jovanovich, Inc., 1968, pp. 16, 18, 22-24, 32.

such a name is unlikely to have been used if it had been common knowledge that Soma-plants did not normally grow in river valleys."[1]

Brough also criticized Wasson's assumption, based on an *interpretation* of a single line of the *Ṛgveda*, that the original Soma-plant was unknown to the priests following the ègvedic period: "He excludes later texts, on the grounds that the original plant had already been replaced by substitutes, and was possibly in process of being lost even by the time of the later hymns of the RV: 10.85.3...'The Soma that the Brahmans know—that no one drinks'."[2] Wasson refers to this verse and writes: "I suggest that the complete abandonment of Soma took place shortly before the last batch of hymns was added to the canon." According to Wasson, "The corpus of post-Vedic exegesis is irrelevant, compiled as it was by persons ignorant of the botany of the plant," and referring to the *Brāhmaṇa* and *Śrauta-Sūtra* texts: "The priests responsible for them knew not the Soma plant, or pretended not to know it."[3] Wasson's opinion and interpretation of this Vedic literature follows his co-author Wendy Doniger O'Flaherty's interpretation of these texts, best described as a distorted representation of the facts, and she also concludes that the *Brāhmaṇas* "contain no passages about the authentic Soma of sure evidential value." She would later reiterate this inaccurate view and revise her interpretation: "it did not really matter what Soma was, since it was lost so early in history... there may never have been an original Soma plant at all...*all* of Soma's 'substitutes'...were surrogates for a mythical plant that never existed save in the minds of the priests."[4]

In his chapter entitled "Soma Was Not Alcoholic," Wasson writes: "In the West it has been repeatedly suggested that Soma was an alcoholic drink...This can be denied with assurance. The difference in tone between the bibulous verse of the West and the holy rapture

[1] John Brough, "Soma and *Amanita muscaria*," *Bulletin of the School of Oriental and African Studies*, 34 (1971), pp. 342-343.

[2] Ibid., p. 333.

[3] R. Gordon Wasson, "*Soma*: Comments Inspired by Professor Kuiper's Review," *Indo-Iranian Journal*, 12 (1970), pp. 291, 296.

[4] Wendy Doniger, "The Post-Vedic History of the Soma Plant," in Wasson, *Soma*, pp. 96-98; Doniger, "'Somatic' Memories of R. Gordon Wasson," *The Sacred Mushroom Seeker: Essays for R. Gordon Wasson*, T. J. Riedlinger (ed.), Portland, Or., Dioscorides Press, 1990, p. 58.

of the Soma hymns will suffice for those of any literary discrimination or psychological insight...The stalks were pressed as a liturgical act and before the liturgy was finished the juice was drunk. Three sacramental offerings could be made in one day. Even if we allow for the heat of the summer in the Indus Valley, fermentation could not have advanced far in a religious rite repeated thrice in a day. Moreover, those who know the fermentation process must find it hard, indeed impossible, to imagine anyone, no matter how far removed from our culture, waxing lyrical, even ecstatic, over a drink in active fermentation. It is not as though the Indo-Aryans were unfamiliar with fermented drinks. They had their *súrā*, mentioned several times in the ṚgVeda but without reverential pæans: quite the contrary...A mere difference between fermented drinks would not cause the gulf that separates *súrā* from the Soma that inspired the great hymns." Wasson also writes: "It seemed to me apparent that all alcoholic drinks could be eliminated at once. This was not only because of the time element in completing the fermentation process: the descriptions of the effects of Soma could never have been written after drinking alcohol. Soma must be an hallucinogen."[1]

To resolve another contradiction in his *Amanita* theory, Wasson was forced to reject "mead" as a definition of the Vedic term *mádhu*, and interpreted this word as strictly meaning ordinary honey: "But no one knows whether the ancestors of the Aryans had been drinking mead: many peoples knowing honey do not ferment it...Honey, *mádhu*, is mentioned frequently in the ṚgVeda but mead never. Honey is cited for its sweetness and also is often applied as a metaphor of enhancement to Soma. There is reason to think it was used on occasions to mix with Soma, but the two were never confused." Following this distinction made between Soma and *mádhu*, which Wasson asserts, "were never confused," he then contradicts this definition of terms: "The word for Soma is *mádhu*, 'honey', a frequent metaphor for Soma." Brough writes: "it is slightly odd to find Wasson writing (p. 16), 'Honey,

[1] R. Gordon Wasson, "Soma of the Aryans: an ancient hallucinogen?" *Bulletin on Narcotics*, 22 (1970), p. 26; Wasson, *Soma*, pp. 15, 176; Wasson, "The Soma of the Rig Veda: What Was It?" *Journal of the American Oriental Society*, 91 (1971), p. 169.

mádhu, is mentioned frequently in the ṚgVeda but mead never'. When the words are etymologically the same, how can one draw such a distinction ?" Wasson responds: "Nothing authorizes us to say the *mádhu* was mead," adding: "I feel certain *mádhu* in the ṚgVeda never meant 'mead' but always 'honey'...Who would downgrade Soma by calling it 'mead?'...If *mádhu* had meant 'mead', why was not *súrā* an equally common trope for Soma?...At the expense of repetition I repeat that a fermented beverage must not be confused with a divine inebriant. This is the first and great commandment for the Western student of Soma to master."[1]

In her summary of the Soma question submitted to Wasson in early 1963 and published in his book, Doniger O'Flaherty perpetrates this inaccuracy with a highly dubious interpretation of a line from the *Taittirīya-Brāhmaṇa* (I.3.3.2): "Soma is male and *súrā* is female; the two make a pair," she concludes: "The sharp distinction made by the text seems to rule out the possibility that Soma was simply another kind of alcoholic drink..." Doniger O'Flaherty also comments on "the inappropriateness of the identification of Soma with alcohol," and in her opinion: "Fairly convincing evidence that Soma was not an alcoholic beverage was established quite early..." Doniger would later write: "I was certain that the evidence proved Soma was an entheogen (we called it an hallucinogen then), and that it was not a form of alcohol (as had been theretofore widely believed) but was a drug provoking an ecstasy of a very special kind." Once characterized as a "full-fledged convert" to Wasson's *Amanita* theory, Doniger O'Flaherty would later write: "I did not believe (and still do not) that we had proven Soma definitely *was* a mushroom...I did believe we had established a far more important hypothesis: Soma was a hallucinogen." Aside from the crucial issue of the presence of alcohol in the Soma beverage, it is worth noting that *Amanita muscaria* has never been scientifically classified as an hallucinogen, and replacing this term with "entheogen" does not change this fact.[2]

Referring to the use of *A. muscaria* among the Siberian tribes, who

[1] Wasson, *Soma*, pp. 15-16, 61; Wasson, *Rejoinder to Professor Brough*, pp. 13, 23; Brough, "Soma and *Amanita*," p. 349.

[2] Wendy Doniger O'Flaherty, in R. Gordon Wasson, "The Last Meal of Buddha," *Journal of the American Oriental Society*, 102 (1982), p. 603; Doniger O'Flaherty, "History of the Soma Plant," pp. 96, 144; Doniger, "Memories of Wasson," p. 58.

consumed the mushroom after it was dried, Wasson writes: "The mushroom is used in a good many different ways in Siberia, but it is certainly not true that it is fermented into a beverage…" Wasson firmly believed that the Soma-plant was dried before it was pressed, which from the standpoint of chemistry and pharmacology would otherwise eliminate his candidate *A. muscaria*: "The fly-agaric was often, perhaps usually dried up when it was used in the (Soma) sacrifice, and initially it had to be soaked in water, reinflated so to speak." Referring to Wasson's proposed method of drying and re-hydrating the Soma-plant, R. C. Zaehner informed Wasson: "There is no evidence that the early Indo-Iranians did this."[1] Yet in a letter to Hofmann years later, Wasson continued to advocate this method of drying and re-hydrating the Soma-plant as if this procedure was an undisputed fact and a requisite for other Soma candidates: "The crucial test might well be this: when the dried sclerotia of *Claviceps purpurea* is reflated with water and poured out, as was the Aryans' practice, what is the color of the resulting fluid?"[2] The proposed method of drying and re-hydrating the Soma-plant is clearly one more of Wasson's circular arguments formed with extraneous facts, from which he reasons an erroneous conclusion that is presented as evidence for his candidate. In his criticism of Wasson's *Amanita* theory, Brough writes: "Even although dried specimens of the mushroom may have been somewhat tough, even after soaking in water, nowhere in Wasson's 'Exhibits' is there any mention of the Siberian tribes pounding the fly-agaric to extract juice. On the contrary, in Siberia the plant is regularly eaten or swallowed whole, although occasionally it is used to make a decoction or infusion (pp. 234, 253, 260, etc.), or it is added to soups or sauces (p. 324)."[3] Wasson dismissed this observation with another circular argument formed with an extraneous fact: "Brough goes on to say there is no mention in my exhibits of the 'pounding' of the mushroom to extract the juice. No, there is no mention, nor of a Brahman caste singing

[1] R. C. Zaehner, "The fortifying fungus" [Book Review], *The Times Literary Supplement*, 22 May 1969, p. 562.
[2] Wasson to Hofmann, 19 July 1978; in Riedlinger, *op. cit.*, p. 154.
[3] Brough, "Soma and *Amanita*," p. 338.

hymns in the Vedic language."[1]

In a previous publication entitled *The Elixir*, the present author has demonstrated that ergot-infected grain meets the physical parameters for Soma set forth by Wasson, with the exception of his urine-drinking hypothesis connected with the *Ṛgveda*, a practice once found among the Koryaks and other indigenous tribes of Siberia who consumed *A. muscaria* for intoxication, and passed their urine to others, which they drank and also became intoxicated. Wasson writes: "I engaged Wendy Doniger to write a *précis* of the Soma question, and she submitted her report on February 16, 1963...In it she called my special attention to ṚgVeda IX 74[4], where the priests urinate Soma. This had astonished her and left her nonplussed: little did she suspect what it would mean for me, with my Siberian background." Doniger (O'Flaherty) writes: "Actually, in an odd sort of way, it was I, not he, who stumbled serendipitously upon the Soma mushroom, as Gordon acknowledged himself in the book's introductory pages." Wasson refers to the urine hypothesis as the "crucial argument in my case," which he based upon the *interpretation* of one sentence, acknowledged to be "a phrase not met with before and not to be met again:" "The swollen men piss the flowing (Soma)" (*Ṛgveda* IX.74.4). Wasson writes: "I take these words at their face value," and he asserts that "The priests appointed to impersonate Indra and Vāyu, having imbibed the Soma mixed with milk or curds, are now urinating Soma."[2]

Daniel H. H. Ingalls disputed Wasson's urine hypothesis: "Where Wasson errs is in supposing that the Vedic soma was drunk in the same way (as the Siberian urine). To justify such a thesis he is forced to suppose that Vedic priests impersonated their gods: that when the text says, 'I offer soma to you, Indra; drink of the good soma,' someone was offering amanita juice to a priest. Actually, there is no shred of evidence for priestly impersonations in the Rigveda. Where priests do act *in persona dei* (as they do in some forms of Hinduism and Buddhism) the procedure is clearly revealed by the language of the ritual and litany. Wasson finds one out of the 35,000 lines

[1] Wasson, *Rejoinder to Professor Brough*, p. 14.

[2] William Scott Shelley, *The Elixir: An Alchemical Study of the Ergot Mushrooms*, Notre Dame, Ind., Cross Cultural Publications, Inc., 1995, pp. 13-29; Wasson, *Soma*, pp. 4, 27, 324; Doniger, "Memories of Wasson," p. 56.

of the Rigveda that seems to say the priests are micturating diluted soma. I interpret the line...to mean that bearers of the soma pots are pouring the fluid down onto the filter-covered trough. One cannot hang the explanation of a major cult on a single image, which may be metaphorical, taken out of context." Ingalls also writes: "But there is no evidence to my knowledge in the whole of the Rigveda that priests ever impersonate the gods in any capacity...Only once in the ten thousand verses of the Rigveda are priests said to piss Soma...the image, taken in the context of Book 9 and of this hymn in particular, is surely metaphorical...Out of well over a thousand references to Soma in the Rigveda there are only these two or three instances in which it is combined with the word to urinate. Surely if in Vedic times the metabolic form of Amanita had had the importance that Wasson claims, we should have more such combinations...the paucity of reference to urine and the total lack of evidence for divine impersonation in the Rigveda are facts. It is not speculation but common sense that forces me to say that Wasson has failed in his attempt to find evidence of a urinary form of the hallucinogen in the Vedic texts."[1]

Brough also refuted Wasson's Soma-urine hypothesis: "Geldner's conjecture that 9.74.4 refers in the first place to the priests pissing in the ritual, though not impossible, seems highly improbable...There is no need to see 'Soma urine' in Wasson's sense...only the *soma*-juice itself flowing into the sacrificial vessels, poetically conceived as pouring down of fertilizing rain from heaven."[2] Flattery and Schwartz contend that "Even interpreting this literally...there is still nothing to suggest the *drinking* of such urine...the passage has no connection with drinking at all, either of urine or of anything else...none of the data presented by Wasson on the subject of urine drinking has any relevance for *soma*."[3]

Referring to the hypothetical Soma-urine, Wasson writes in *Soma*: "We do not know how the metabolite was taken, whether neat or with water or milk

---

[1] Daniel H. H. Ingalls, "Remarks on Mr. Wasson's *Soma*," *Journal of the American Oriental Society*, 91 (1971), pp. 189-190; Ingalls, "Soma" [Book Review], *The New York Times Book Review*, 5 September 1971, p. 15.

[2] John Brough, "Problems of the «Soma-Mushroom» Theory," *Indologica Taurinensia*, 1 (1973), p. 25.

[3] David Stophlet Flattery and Martin Schwartz, *Haoma and Harmaline: The Botanical Identity of the Indo-Iranian Sacred Hallucinogen "Soma" and its Legacy in Religion, Language, and Middle Eastern Folklore*, Berkeley, University of California Press, 1989, p. 6.

or curds or honey." Defending his hypothesis, Wasson writes: "The urine-drinking custom is strong among the Northeastern tribes in the Chukotka... It is the thesis of my book that a practice...was shared in the remote past by the Indo-Aryans who brought it down from their northern homeland," adding: "the ordinary urine of the (Soma) drinker becomes impregnated with the god and then in turn is sometimes imbibed." In a manner characteristic of his modus operandi in *Soma*, Wasson again defends his hypothesis with a circular argument formed with an extraneous fact, from which he reasons an erroneous conclusion that is presented as evidence for his case: "I postulate the drinking of Soma-urine in the *Ṛg Veda*, analogous to the drinking of Soma-urine (specifically, *A. muscaria*-urine) by the tribesmen of the Chukotka and Kamchatka in our own time...Only with Soma is there Soma-urine and how did the priests learn this other than by drinking the urine?" But in defending his hypothesis Wasson would also contradict these statements: "I do not assert that the Vedic priests drank Soma-urine."[1]

Beyond Wasson's *interpretation* of urine-drinking in the *Ṛgveda*, there is very little evidence for this practice in the Vedic literature that followed this text. Brough writes: "He therefore devotes a major part of his book to the discussion of the use of the fly-agaric and other mushrooms in lands beyond India and Iran, and especially in Siberia. These sections of the books are full of interest as an aspect of cultural history; but as a matter of simple logic, such parallels have no probative value...the non-Indo-Iranian materials remain, in the strictest sense, irrelevant...extraneous facts are not additional evidence." Brough would later add: "But now 'Soma-urine' appears to have been taken as a reality in its own right, and on this fictitious entity a specious argument is based. It can be said with confidence that no Vedic expression exists which could be translated as 'Soma-urine'."[2] And finally, Ilya Gershevitch writes: "If the original sauma was Amanita muscaria, its Indo-Iranian devotees would necessarily know of its effect on urine. We have seen that no such knowledge can be demonstrated from any Vedic or Avestan verse."[3]

[1] R. Gordon Wasson, "Soma: the Divine Mushroom of Immortality," *Discovery*, 3 (1967), p. 46; Wasson, *Soma*, pp. 25, 29-31, 176; Wasson, "Professor Kuiper's Review," pp. 291-292; Wasson, "Soma of the Rig Veda," p. 179; Wasson, *Rejoinder to Professor Brough*, pp. 24-26, 40.

[2] Brough, "Soma and *Amanita*," pp. 332-333, 343-348, 359.

[3] Gershevitch, "Soma Controversy," p. 57.

Chapter 7: The Indo-Aryans of Central Asia

In recent years in the territory of Greater Iran, archaeologists have discovered the remains of the ancient civilizations of Margiana in southeastern Turkmenistan, and Bactria in northern Afghanistan. These excavations in Central Asia reveal a cultural complex of highly developed agricultural societies that engaged in the manufacturing and ritual consumption of the Soma/Haoma libation. The Bactria-Margiana Archaeological Complex (BMAC) represents the cultures and people who first appeared on the Borderlands of the Indus Valley around the close of the third millennium B.C., a people that referred to themselves as *āryas*, whose Soma-sacrificing priests would compose the *Ṛgveda*.[1] Frederik T. Hiebert writes:

> In Central Asia the period from 2000–1750 B.C. was a time of tremendous change. The oasis colonies in the desert of Margiana, first formed from the "late Namazga V" foothill culture, gave rise to a distinctive culture, known as the Bactria-Margiana Archaeological Complex, or "BMAC"...on the western borders of the Indus Valley...[T]he burials from Quetta...and Sibri...Cenotaphs found in Mehrgarh VIII and at Sibri are also typical of the BMAC period in Margiana and Bactria and together with the Central Asian burials indicate the presence of Central Asian people on the edge of the

[1] George Erdosy, "Language, material culture and ethnicity: Theoretical perspectives," *The Indo-Aryans of Ancient South Asia: Language, Material Culture and Ethnicity*, G. Erdosy (ed.), Berlin, Walter de Gruyter, 1995, p. 14.

> Indus Valley. These assemblages at Mehrgarh VIII, Sibri and Quetta date to 1900-1700 B.C. on the basis of the Central Asian finds. The process of expansion from Central Asia – an organised military venture, perhaps? — remains unclear...early oasis adaptation appears first in Margiana, but during 1900–1700 B.C. we see its expansion from the oases of both Bactria and Margiana, indicating the movement of people to the periphery of the Indus Valley, and of artifacts even further.[1]

The slow decline of the Harappan civilization coincided with the arrival of these immigrants from Central Asia, and although in Harappa there are signs of exaggerated attention paid to the city's defense suggesting some foreign danger,[2] the collapse of this civilization was unlikely the result of a cataclysmic military invasion by the Indo-Aryans. During this period of decline, which occurred over centuries, it also seems unlikely that the intruding Indo-Aryans and Harappans co-existed without conflict. Asko Parpola writes:

> The site of Pirak, which continues the sequence of Mehrgarh, Sibri and Nausharo in the Kachi plain in Pakistan, from c. 1800 B.C. testifies to the rapid diffusion of the horse and the two humped Bactrian camel in northwest India during the first quarter of the second millennium B.C...It is obvious that Sind served as a channel through which immigrants representing first the Namazga V and shortly thereafter also the Namazga VI culture continued to other parts of the Indian subcontinent...The arrival of the Namazga V people seems to have disrupted the political and cultural unity of the Indus valley soon after 2000 B.C. The urban system of the Harappans and the process of city life, such as centralized government with the collection of taxes and organization of trade, ceased to function. The thousands of countryside villages, however, persisted. In peripheral regions, especially in Gujarat, Mature Harappan traits, mixed with new elements, lingered longer, until c. 1750 B.C. The newcomers

[1] Frederik T. Hiebert, "South Asia from a Central Asian perspective," *The Indo-Aryans of Ancient South Asia: Language, Material Culture and Ethnicity*, G. Erdosy (ed.), Berlin, Walter de Gruyter, 1995, pp. 199-202.

[2] Victor I. Sarianidi, *Margiana and Protozoroastrism*, Inna Sarianidi (trans.), Athens, Kapon Editions, 1998, pp. 150-152.

> did not stop in the Harappan area, however, but pushed on further both into the Deccan and towards the Gangetic valley.[1]

Thomas Burrow writes in favor of an Aryan invasion of India, similar to the Anglo-Saxon invasion of Britain:

> From the *Ṛgveda* emerges a fairly clear picture of the situation at that time. A series of related tribes, settled mainly in the Panjāb and adjacent regions, speaking a common language, sharing a common religion, and designating themselves by the name *ārya*, are represented as being in a state of permanent conflict with a hostile group of peoples known variously as Dāsa or Dasyu. From the frequent references to these conflicts it emerges that their result was the complete victory of the Āryans...references are frequent to the struggle with the previous inhabitants, the Dāsas or Dasyus, and to the occupation of their land and the capture of their possessions. As to the identity of these people who were displaced or subjugated,[2] the predominant and the most likely view is that they were the authors of the Indus civilization...there has been some argument as to whether its fall was brought about by the invading Āryans, or whether some period of time elapsed between the end of the Indus civilization and the appearance of the Āryans. The evidence of the Vedic texts themselves is decidedly in favour of the former view, notably on account of the frequent references to the destruction of cities, the war-god Indra being known as *puraṃdara*, "destroyer of cities". Agni, the fire-god is also prominently mentioned in this capacity, understandably, since many of the Indus cities appear to have been destroyed by fire.[3]

The Indo-Aryans of the *Ṛgveda* consisted of various tribes that settled the region prior to the drying up of the Sarasvatī River and the flooding of some

[1] Asko Parpola, "The Coming of the Aryans to Iran and India and the Cultural and Ethnic Identity of the Dāsas," *Studia Orientalia*, 64 (1988), p. 206.

[2] For an alternative view, see Ibid., pp. 195-302. For criticism of Parpola's theory, see Kenneth R. Norman, Review of "The Coming of the Aryans to Iran and India and the Cultural and Ethnic Identity of the Dāsas," *Acta Orientalia*, 51 (1990), pp. 288-296.

[3] Thomas Burrow, "The Early Āryans," *A Cultural History of India*, A. L. Basham (ed.), Oxford, Clarendon Press, 1975, pp. 20-25.

of the Harappan settlements by the Indus River,[1] and these tribes were named Yadu-Turvaśa, Anu-Druhyu, and Pūru-Bharatas. Parpola writes, "In the Ṛgveda many ethnic names are mentioned; they are, however, divided into two major antagonistic groups, the 'five clans' of the Aryans (Yadus, Anus, Druhyus, Turvaśas, and Pūrus) and dark-skinned inimical peoples, with whom the Ṛgvedic Aryans fought for the possession of cattle and pasturage. These hostile and hated people are mostly called Dasyus, Dāsas, or Paṇis."[2] Michael Witzel writes:

> Once they arrived on the plains of the Panjab, the Indo-Aryans had further battles to fight...the Indo-Aryans fought each other as often as they fought non-Indo-Aryans...In book 4, the Yadu-Turvaśa appear only once...In book 6, however, they are more prominent, being at times friends and at times enemies of the Pūru-Bharatas...they are frequently associated with the Anu, Druhyu and Pūru, thus making up the "Five Peoples"...the Yadu-Turvaśa (and the Anu-Druhyu) are regarded as settled in the Panjab at the time of the arrival of the Pūrus and Bharatas...The Pūru appear to be a broad conglomerate of tribes, to which at one time the Bharatas also belonged...they (along with the Bharatas) occupy centre-stage in much of the ṚgVeda, succeeding earlier groups of migrants such as the Turvaśa and the Yadu...By the time of the composition of most Ṛgvedic hymns, the Yadu-Turvaśa and the Anu-Druhyu had already been well established in the Panjab...Books 3 and — particularly 7 detail the ultimate victory of the Bharatas over the other tribes (eventually including the Pūrus...) and their settlement on the Sarasvatī, which became the heartland of South Asia well into the Vedic period.[3]

[1] V. N. Misra, "Climate, a Factor in the Rise and Fall of the Indus Civilization – Evidence from Rajasthan and Beyond," *Frontiers of the Indus Civilization*, B. B. Lal and S. P. Gupta (eds.), New Delhi, Indian Archaeological Society, 1984, pp. 475-481; Louis Flam, "Fluvial Geomorphology of the Lower Indus Basin (Sindh, Pakistan) and the Indus Civilization," *Himalaya to the Sea: Geology, geomorphology and the Quaternary*, John F. Shroder, Jr. (ed.), London, Routledge, 1993, pp. 265-287.

[2] Parpola, "The Coming of the Aryans," p. 208.

[3] Michael Witzel, "Ṛgvedic history: poets, chieftains and polities," *The Indo-Aryans of Ancient South Asia: Language, Material Culture and Ethnicity*, G. Erdosy (ed.), Berlin, Walter de Gruyter, 1995, pp. 324-339.

The Ṛgvedic Aryans occupied the territory of the Indus and the Panjāb, extending north into present day Afghanistan, and east to the rivers Sarasvatī and Yamunā, and there is a late hymn that mentions the Gangā. Burrow writes, "Once the Indus civilization had been overthrown, the greater part of its territory occupied, there remained no advanced civilized states to contend with. The Gangā valley seems at this time to have been thinly populated by forest tribes...as the activity of forest-clearing proceeded...sections of them... attached themselves to the fringe of Āryan society, forming the nucleus of what were to become eventually the depressed classes."[4]

George Erdosy writes, "Indo-Aryan languages first emerged in the Borderlands during the dissolution of an urban civilisation. Such a sudden collapse as seems to have characterised the Localisation Era — abandonment of urban centres, large-scale relocation of population due to the drying up of the Sarasvatī River, loss of foreign contacts, decline in arts and crafts — could not but have a negative impact on the ideology of the Harappans."[5] Jim G. Shaffer and Diane A. Lichtenstein write, "During the mid-second millennium B.C. many, but not all, Indus Valley settlements, including urban centers, were abandoned...This shift by Harappan and, perhaps, other Indus Valley cultural mosaic groups, is the only archaeologically documented west-to-east movement of human populations in South Asia before the first half of the first millennium B.C...The reasons for this population movement remain unknown."[6] Edmund Leach writes, "The Indus civilization did not come to an end suddenly but over a period of centuries. The primary cause of its decline was probably a geological catastrophe that led to a change in the course of the Indus, a failure of the irrigation system, and the collapse of waterborne trade. In the centuries that followed, the political and economic

[4] "The area occupied by the Āryans continued to expand in the period represented by the later Vedic texts, and there was a shift eastwards in the centre of gravity. By the time of the *Brāhmaṇas*...the western settlements in the Panjāb were less important...The rapid expansion during the period 800-550 B.C. had the result that in the new territories the Aryans were much more thinly spread than in the old, and they were to a greater extent mixed with the pre-existing population... The influence of Āryan civilization was felt latest in the Dravidian south." Burrow, *op. cit.*, pp. 26-29.

[5] Erdosy, "Language, material culture and ethnicity," p. 19.

[6] Jim G. Shaffer and Diane A. Lichtenstein, "The concepts of 'cultural tradition' and 'palaeoethnicity' in South Asian archaeology," *The Indo-Aryans of Ancient South Asia: Language, Material Culture and Ethnicity*, G. Erdosy (ed.), Berlin, Walter de Gruyter, 1995, pp. 138-139.

center of gravity moved eastward into the Ganges plain."[1] Jonathan Mark Kenoyer writes:

> Due to sedimentation and tectonic movement, the Ghaggar-Hakra system (Sarasvatī River) was captured by the River Sutlej of the Indus system and the River Yamunā of the Gangetic system... The Indus itself began to swing east, flooding many settlements in the process...Sites such as Harappa continued to be inhabited and are still important cities today. However, many less fortunate settlements along the dry bed of the Ghaggar-Hakra system were abandoned and their inhabitants were forced to develop new subsistence strategies or move to more stable agricultural regions.[2]

This viewpoint is supported by the account of Strabo, who cites Aristobules who visited the Indus Valley in 326 B.C., and found there "a country of more than a thousand cities, together with villages, that had been deserted because the Indus had abandoned its proper bed, and had turned aside into the other bed on the left that was much deeper, and flowed with precipitous descent like a cataract, so that the Indus no longer watered by its overflows the abandoned country on the right, since that country was now above the level, not only of the new stream, but also of its overflows.[3]

However, the Harappan civilization was primarily located on the Sarasvatī and its tributaries, and this territory was also occupied by the Indo-Aryans prior to this river system becoming extinct and the subsequent abandonment of the Harappan settlements. V. N. Misra writes, "The densest distribution of Harappan sites is not on the Indus river and its tributaries but on the extinct Hakra-Ghaggar and its equally extinct tributaries. Of the over 800 Harappan sites...more than 530 are located on the Hakra-Ghaggar system...The Harappan culture is, therefore, essentially a culture of the Hakra-Ghaggar valley...the Sarasvati was a fully flowing river in the second millennium B.C...but had dried up by c. 1000 B.C."[4] Michael Witzel writes:

[1] Edmund Leach, "Aryan Invasions over Four Millennia," *Culture Through Time: Anthropological Approaches*, Emiko Ohnuki-Tierney (ed.), Stanford, Stanford University Press, 1990, p. 238.

[2] Jonathan Mark Kenoyer, "Interaction systems, specialised crafts and culture change: The Indus Valley Tradition and the Indo-Gangetic Tradition in South Asia," *The Indo-Aryans of Ancient South Asia: Language, Material Culture and Ethnicity*, G. Erdosy (ed.), Berlin, Walter de Gruyter, 1995, p. 224.

[3] Strabo, *Geographica* XV.1.19.

[4] Misra, *op. cit.*, pp. 473-475.

> [A]bsolute dates are difficult to establish...the archaeologically attested appearance of iron which forms a date *post quem* for the *mantra* portions of the Atharvaveda around 1150 B.C...we may now add as a date *post quem*, the end of the urban phase...of the Indus Civilization around 1900 B.C., an event that must precede the Vedic texts which do not know of cities or towns but speak, instead, of ruined places...At the same time, since the Sarasvatī, which dries up progressively after the mid-2nd millennium B.C...is still described as a mighty stream in the ṚgVeda, the earliest hymns in the latter must have been composed by c. 1500 B.C.[1]

Victor I. Sarianidi has concluded that the archaeological evidence suggests a western origin for the Indo-Aryans, traceable to Syria and Mesopotamia, and writes, "Although we do not as yet accurately know where these tribes came from, their general western origin cannot be doubted. It is not excluded that, departing from their Syro-Anatolian metropolis already in remote antiquity, they lingered in some interim territory for many centuries to appear in Central Asia and the Indian subcontinent only in the second millennium B.C."[2] Sarianidi has also suggested that these tribes from the Syro-Anatolian region and immigrants from South Turkmenistan founded Margiana, and those who settled in Bactria were exclusively Syro-Anatolian.[3] Kenneth R. Norman writes, "the Indo-Iranians had reached an area to the north of present-day Iran not later than 2000 B.C...They then split into two. One section, the Iranians, remained in the region of the River Oxus. The other section, the Indo-Aryans, began to move south. Some of them moved down into Iran and then to the west...Other Indo-Aryans moved eastwards and began to enter India."[4]

George Erdosy writes, "[W]e are still left with remarkable linguistic

[1] Michael Witzel, "Early Indian history: Linguistic and textual parametres," *The Indo-Aryans of Ancient South Asia: Language, Material Culture and Ethnicity*, G. Erdosy (ed.), Berlin, Walter de Gruyter, 1995, pp. 97-98.

[2] Victor I. Sarianidi, "Margiana and the Indo-Iranian world," *South Asian Archaeology 1993: Proceedings of the Twelfth International Conference of the European Association of South Asian Archaeologists held in Helsinki University 5-9 July 1993*, Volumes I-II, A. Parpola and P. Koskikallio (eds.), Helsinki, Soumalainen Tiedeakatemia, 1994, vol. II, p. 678.

[3] Sarianidi, *Margiana and Protozoroastrism*, pp. 147-148.

[4] Kenneth R. Norman, "Dialect variation in Old and Middle Indo-Aryan," *The Indo-Aryans of Ancient South Asia: Language, Material Culture and Ethnicity*, G. Erdosy (ed.), Berlin, Walter de Gruyter, 1995, p. 278.

similarities between the language of the Aryas and those spoken on the Iranian plateau and in the classical world. These go beyond a shared vocabulary, and involve grammar and syntax to such a degree that linguists to this day cannot offer a convincing alternative to their descent from a common ancestor."[1] P. Oktor Skjærvø writes:

> Evidence either for the history of the Iranian tribes or their languages from the period following the separation of the Indian and Iranian tribes down to the early 1st millennium B.C. is sadly lacking. There are no written sources…The earliest mention of Iranians in historical sources is, paradoxically, of those settled on the Iranian plateau, not those still in central Asia, their ancestral homeland. "Persians" are first mentioned in the 9th-century B.C. Assyrian annals…There are no literary sources for Iranians in Central Asia before the Old Persian inscriptions (Darius's Bisotun inscription, 521–519 B.C., ed. Schmitt) and Herodotus' *Histories* (ca. 470 B.C.). These show that by the mid-1st millennium B.C. tribes called Sakas by the Persians and Scythians by the Greeks were spread throughout Central Asia, from the westernmost edges (north and northwest of the Black Sea) to its easternmost borders. Thus Darius, in his inscriptions, mentions the "Sakas beyond the sea" (probably north of the Black Sea), the "Sakā Haumavargā (somewhere north of modern Afghanistan), and the "pointed-hat Sakas"…the latter two are probably those who are referred to as "beyond the Sogdians"…Many of the tribal names given by Herodotus typically contain the suffix *–tai* (Massagetai, Paralatai, Sauromatai), which is the common plural ending in Middle Iranian Sogdian and modern Ossetic.[2]

Herodotus states that the Persians call all the Scythians "Sakas."[3] Along with other Scythian tribes, the Massagatae or "Big Saka" also populated Margiana, and in the fifth century A.D., Attila the Hun invaded the Roman domain with what Procopius describes as "a great army of Massagetae and

[1] George Erdosy, "Ethnicity in the Rigveda and its Bearing on the Question of Indo-European Origins," *South Asian Studies*, 5 (1989), p. 44.

[2] P. Oktor Skjærvø, "The Avesta as source for the early history of the Iranians," *The Indo-Aryans of Ancient South Asia: Language, Material Culture and Ethnicity*, G. Erdosy (ed.), Berlin, Walter de Gruyter, 1995, pp. 155-157.

[3] Herodotus, *Histories* VII.64.

other Scythians…"[1] Victor I. Sarianidi writes:

> The consistency with which these ceramics of the nomadic tribes are found in the temples of Margiana, associated with the preparation of the hallucinogenic beverage or juices of the *haoma/soma* type supports an old theory expressed almost one hundred years ago by the Russian orientalist V. Grigoriev and revised later by academician V. V. Struve. Both of these scholars paid attention to a statement by Ptolemy on the heterogeneity of the populations of Margiana, which included the Massagatai or "Big Saka." Thus, in an earlier period, the Eastern Saka and "Amyrgioi" Saka penetrated this area, and both scholars applied the name "Amyrgioi" to the Murgab River. On the other hand, referring to Saka Amyrgioi, in ancient Persian writing the words "saka haumavarga" were used, that is, the "*haoma* manufacturing" people.[2]

The ancient German tribes worshipped Wotan (Wodan, Odin), who was intimately connected with the "mead of the gods" in Germanic religion,[3] and Wodan has been compared to the Soma-drinking Vedic deity Indra,[4] and with the bull-slaying Zoroastrian deity Mithra, the god of the popular mystery cult of Roman Mithraism that extended throughout the Roman Empire in Europe.[5] H. R. Ellis Davidson writes:

> In Germany as elsewhere Mithras is depicted…with a raven messenger…In Germanic iconography the raven was the close companion of Wodan and of the later Odin…The Germanic god of battle might also be represented by the eagle…one of his functions in Indo-Iranian tradition appears to have been to bring the sacred drink, *soma* or *haoma*, from the cloud rock…The Viking Age stones in Gotland appear to depict Odin in eagle form, and in one scene

[1] Procopius, *History of the Wars* III.4.24.

[2] Viktor I. Sarianidi, Preface to Fredrik T. Hiebert, *Origins of the Bronze Age Oasis Civilization in Central Asia*, Cambridge, Peabody Museum of Archaeology and Ethnology, Harvard University, 1994, p. xxxiv.

[3] H. R. Ellis Davidson, *Myths and symbols in pagan Europe: Early Scandinavian and Celtic religions*, Manchester, Manchester University Press, 1988, pp. 44-45, 175-176.

[4] John Rhys, *Lectures on the Origin and Growth of Religion as Illustrated by Celtic Heathendom*, Oxford, Williams and Norgate, 1898, pp. 292-301.

[5] L. A. Campbell, *Mithraic Iconography and Ideology*, Leiden, E. J. Brill, 1968, p. 217; H. R. Ellis Davidson, "Mithras and Wodan," *Etudes Mithraiques*, 4 (1978), pp. 99-110.

> he brings back the magic mead to the gods in his eagle shape, as recounted in the literary myths…The symbol of the feast and the drinking of the mead of inspiration was clearly of considerable importance in the cult of the Germanic god, and also in the picture of the Scandinavian Odin in the Viking Age…The mead was recovered by the god himself in eagle form from the rock where the giants had hidden it…[1]

The mead of Odin[2] was intimately connected with the Yggdrasil,[3] the "world tree" on which Odin hangs. Paul C. Bauschatz writes:

> In *Gylfaginning* 15…Yggdrasil has three separate roots extending into three separate wells (Hvergelmir, Mímir's Well, and Urth's Well)…probably the three names all apply to one well, which was basically…the source of wisdom…the well and the tree together are linked…as sources of wisdom…Little is known about the third

[1] Davidson, "Mithras and Wodan," pp. 101-107.

[2] "The most obvious feature of the Bell Beaker complex is its eponymous vessel, the decorated drinking cup…The appearance of these assemblages marks the beginning of a feature which has characterised European culture ever since: an emphasis on male drinking rituals…Although more that one type of drink must have been involved, it can be argued that alcohol was a crucial common element in the expansion of a new cultural pattern across Europe in the Third Millennium… Modern brewers prefer the hulled form of the two-row barley (*Hordeum distichum*); but there was no two-row barley in prehistoric Europe, only the six-row form (*H. hexastichum*); and the naked form – in contrast to the Near East at this time – was the predominant one down to the Urnfield period. Since the naked form has no advantage for brewing, it seems unlikely that its major use was for beer… There are some indications, in fact, that emmer was the cereal used for brewing in northern Europe…Honey was the most concentrated source of sugar in prehistoric temperature [*sic*] Europe, and no doubt a valuable and much-traded commodity…Quite how much was available, however, is an important question… The supply of honey in temperate Europe is not likely to have sustained a mass production of mead, which even in the Middle Ages was four times as valuable as ale." Andrew G. Sherratt, "Cups That Cheered," *Bell Beakers of the Western Mediterranean: Definition, Interpretation, Theory and New Site Data*, William H. Waldren and Rex Claire Kennard (eds.), Oxford, B.A.R., 1987, pp. 81-82, 94-95.

[3] "Another semantic field with very good attestation is that of 'honey'. The noun **melit* is found widely in the West and Centre (e.g. [Old Irish] *mil* 'honey', [Latin] *mel* 'honey', [Modern English] *mildew* [< 'sweet sap'], [Albanian] *bletë* 'honey-bee', [Greek] *méli* 'honey', *mélissa* 'honey-bee', [Armenian] *melr* 'honey', incuding Anatolian, e.g. [Hittite] *militt-* 'honey') and has one Iranian cognate in the form of a reference to *melition*, a drink of the Scythians. The fermented drink made from honey, 'mead' is *médhu* ('alcoholic drink')…There is archaeological evidence for mead from the third millennium BC but it may be considerably older…The proliferation of drinking cups that is seen in central and eastern Europe about 3500 BC has been associated with the spread of alcoholic beverages and, possibly, special drinking cults." Mallory and Adams, *op. cit.*, pp. 262, 265.

well, Hvergelmir. Its name is usually rendered as Roaring Kettle or Seething Cauldron…In *Grímnismál* 25-26 (62), certain activities of some of the many animals associated with the world tree involve Hvergelmir…the world tree…is said to include the mead-hall, Valholl; its mead is supplied by the goat Heithrún…the clear mead flows from the hall over the horns of the feeding hart [Eikthyrnir]… and finds its way eventually down to Hvergelmir…the dews that fall from the branches of Yggdrasil find their way down into the collecting basin of Urth's Well…The mead that Heithrún supplies to the drinkers in the hall is related…to the drops that fall from the horns of Eikthyrnir into Hvergelmir. The dropping liquid is called 'dews' (*doggvar*) in *Voluspá* 19, and the elaboration of the passage in *Gylfaginning* 16 further identifies it as 'honeydew' (*hunangfall*). The dew/honeydew/mead relationship is clear.[1]

[1] Paul C. Bauschatz, *The Well and the Tree: World and Time in Early Germanic Culture*, Amherst, University of Massachusetts Press, 1982, pp. 22-24, 203 n. 21.

CHAPTER 8: THE RELIGION OF THE INDO-ARYANS

Zoroastrianism emerged from a society of "Iranian paganism," and there is evidence that the religion of the Prophet Zarathustra, or Zoroaster,[1] was a reform of the "Iranian paganism" practiced in the temples of Margiana.[2] M. J. Dresden writes:

> A substantial body of linguistic, religious, and social evidence warrants the assumption that, at one time, the bearers of two cultures, which find their expression in the Indian *Rigveda* on the one hand and in parts of the Iranian *Avesta* on the other, formed a unity...Evidence for the emergence of the Indo-Iranian type of language is found as early as 1800 B.C. in the Kassite word *šuriaš* (Indian *sūryas*) for the sun god and, further, in a number of proper names from Anatolia of about the same time. Some four centuries later, around 1400 B.C., the names of four divinities, *Mitra, Varuṇa,*

[1] "Zoroaster, or Zarathustra, invoked in his hymns the Saviours who, like the dawns of new days, were to come and save the world...The religion of pre-islamic [*sic*] Iran resulted from the fusion of the Zoroastrian reform with doctrines and customs of the ancient Magi...Under the great dynasty of Cyrus, Darius, etc., they had the monopoly of religion...the story of a journey of Magi to Bethlehem...the belief that the birth of great men was marked by the appearance of a star: such a belief made it likely that the Magi, as astrologers, were capable of foreseeing such an event and had in fact been watching for it...the Iranian Magi, as disciples of Zoroaster, had been expecting a saviour." Jacques Duchesne-Guillemin, "The Wise Men from the East in the Western Tradition," *Papers in Honour of Professor Mary Boyce*, Leiden, E. J. Brill, 1985, vol. I, pp. 150-151.

[2] Sarianidi, *Margiana and Protozoroastrism*, pp. 168-175.

> *Indra*, and the two *Nāsatyas*,[1] all of whom appear in the *Rigveda*, occur in a treaty between a Hittite king and a ruler of the Hurrian kingdom of the Mitanni in northern Mesopotamia...Records for the time in which the Indo-Iranian unity was broken up...are absent... Our main source of information for Iranian mythology and religion is the *Avesta*, a collection of canonical sacred scriptures ascribed to *Zarathuštra*...Not one but two forms of religion are represented in the Avestan writings. The first is reflected in Zarathuštra's own words as expressed in the Gāthās...called *Zarathuštrianism*...In the second form of religion represented in the *Avesta*, which has been termed *Zoroastrianism*...the texts, which were composed by authors who wrote in the centuries following Zarathuštra's death, recommend the cult of both Indo-Iranian divinities such as *Mithra* (Indian *Mitra*) and *Haoma* (Indian *Soma*)...[2]

Sarianidi writes, "the motif of the five-headed hydra, very rare in itself, has been known only in Mesopotamia and now also in Bactria. Significantly in Rigveda god Indra fights against a similar many-headed hydra seeking to steal the water of life...Bactrian-Margiana glyptics is virtually permeated with the dualistic idea of the struggle between good and evil...the evil, negatives forces are represented by serpent-like dragons..."[3] Terracotta figurines of female deities are found in abundance at some sites in Margiana, while at other sites these figurines are completely absent, yet sculptures of deities are not found in the temples of Margiana, which appear to be exclusively devoted to the cults of fire and hallucinogenic libations.[4] Asko Parpola briefly examines the religion of the Bactria-Margiana Archaeological Complex:

> [In] the religion of the BMAC...There is widespread evidence for the worship of a goddess connected with lions...ultimately going back to the traditions of the ancient Near East. Connections with the later Indian worship of Dūrga, the goddess of victory and

[1] "[T]he Hurrian state of Mitanni, to judge by the names of its kings, was, during its most influential period, under the domination of Āryan kings backed up by an Āryan aristocracy. Other minor states in Syria had rulers with similar Āryan names." Burrow, *op. cit.*, p. 23.

[2] M. J. Dresden, "Mythology of Ancient Iran," *Mythologies of the Ancient World*, S. N. Kramer (ed.), New York, Doubleday, 1961, pp. 333-335.

[3] Victor I. Sarianidi, "The Bactrian Pantheon," *International Association for the Study of the Cultures of Central Asia, Information Bulletin*, 10 (1987), pp. 7-11.

[4] Sarianidi, *Margiana and Protozoroastrism*, pp. 33, 132.

> fertility escorted by a lion or tiger, the protectress of the stronghold (*durga*), are suggested by several things…A Bactrian seal depicting copulating pairs, both man and animal, reminds one of the orgies associated with the principal festival of the goddess. Wine is associated with the cult of the goddess and may have been enjoyed from the fabulous drinking cups made of silver and gold found in Bactria and Baluchistan, for viticulture is an integral part of the BMAC…[1]

Parpola writes:

> In the Akkadian seals, Ishtar as the goddess of war is associated with the lion…The later Indian goddess of war and victory, Dūrga, has the tiger or the lion for her vehicle, and she is the main recipient of buffalo sacrifices. In some Purāṇic and modern folk stories, the (slaughtered) buffalo is the husband of the goddess, just like the (dying) bull (the animal of the thunder god) stands for the husband of the goddess in a number of ancient Near Eastern religions… Though the Hindu cult of Dūrga is little attested before Christian times, there is thus reason to assume that it goes back to the third millennium B.C. and is of Near Eastern origin…scholars have shown that the Near Eastern influence came to Bactria mainly from Elam and from Syria. This influence, which was later transmitted to the Indian subcontinent, is manifested in architecture as well as small finds including seals and ritual weapons. There is clear evidence for the worship of a goddess of Near Eastern affinity, associated with lions and eagles or griffins. She appears on several Bactrian seals… the Vedic cult…mainly centered on the preparation of a cultic drink called **Sauma;* *Sauma was offered to Indra, the god of war and thunder, and the head of the Vedic pantheon.[2]

According to Sarianidi, the Margianan temple at Togolok-21 was "used by proto-Zoroastrians whose religious beliefs and rites latter became (in changed form) part of official Zoroastrianism…the religious ideas and

[1] Parpola, "Aryans and the Soma," p. 370.
[2] Asko Parpola, "Bronze Age Bactria and Indian Religion," *Studia Orientalia*, 70 (1993), pp. 82-83.

cult practices of the BMAC tribes continued to exist...in Zoroastrianism."[1] Sarianidi writes:

> Zoroastrianism is known to have originated in an Iranian environment and, more precisely, in a society of "Iranian paganism". It is logical then to assume that the soma-haoma cult appeared in this society and that later Zoroaster included it in his new religion. For a long time searches for "Iranian paganism" were fruitless and only in the last decades the signs of it were found...in Bactria... and especially in Margiana...Archaeological discoveries in Margiana, the country mentioned in the Beihustan script under the name of Margush, have yielded material that pointed to the ritual cult of the intoxicating drink of haoma which took a central place in the religious ideas of local tribes. Most representative are the monumental temples (Togolok-1, Togolok-21, temenos Gonur), their sizes and elaborate principles of the layout easily comparable to the famous temples of Mesopotamia. The Togolok-21 temple can be looked upon as a kind of "cathedral" that served the needs of the whole ancient country of Margush.[2]

[1] Victor I. Sarianidi, "South-west Asia: Migrations, the Aryans and Zoroastrians," *International Association for the Study of the Cultures of Central Asia, Information Bulletin*, 13 (1987), p. 51.

[2] Victor I. Sarianidi, "Margiana and Soma-Haoma," Electronic Journal of Vedic Studies, 9 (2003).

Chapter 9: The "Soma Vessels" of the Temples of Margiana

In the Kara Kum desert of Turkmenistan, archaeologists have uncovered a number of sites dating from the second millennium B.C., originally occupied by agricultural tribes that colonized the fertile delta of the Murgab River. From three of these sites in ancient Margiana were excavated monumental temples (Gonur temenos, Togolok-1, Togolok-21,) that contain the vessels and instruments necessary in preparing the secret cult libation. Gonur-depe is the oldest of these sites, the southern part of this site consisting of a fortified complex of buildings, private dwellings, and a large temple consisting of two parts: one used for public worship, and the other a hidden inner sanctum reserved for the priesthood. In the southern part of this temple, in the hidden sanctuary, there are three large rooms in a straight line that are connected by passages to one another and through another room to the public part of the complex. Three ceramic vessels embedded in a brick platform that contain remains were found in one of these private rooms, and about ten ceramic pot-stands and fragments of ceramic strainers with centrally located holes were found in the adjoining room. In a room located in the northern section of the temple a basin containing remains along with ceramic pot-stands and fragments of conic strainers were also recovered. Similar vessels and earthenware stands were also found at Togoluk-1, probably dating from the mid-second millennium B.C., and at Togoluk-21, dated to the late second millennium B.C. Also found in all three Margianian temples were stone articles connected with grinding, along with smaller vessels with long spouts and vessels with sculptural friezes along the rim. Reminiscent of the rites performed at Eleusis in Greece, weaved baskets containing remains were also uncovered from the temples of Margiana. Sarianidi writes:

> [I]n Turkmenistan...ancient agricultural tribes...first appeared here, in the fertile delta of the Murgab river, at the beginning of the (2nd millennium B.C.)...they colonized the delta, laying the foundations of a settled life based on agriculture...Gonur-depe consists of two parts[:] the northern part which comprises the site itself, with a citadel in its centre...A kind of fortress was built to the south of the settlement....a temple, was erected in the south western corner of the fortress...The southern auxiliary part of the building was reserved for the process of making the ritual drink[.] It has three vast rooms arranged in a straight line...The one in the western corner revealed...three shallow vessels...covered with a thick layer of plaster...a considerable number of particles...were preserved within the plaster, leaving no doubt...particular plants had been prepared in the vessels...similar rooms, also coated with white plaster (conveniently called "white rooms"), with vessels... were found at Togolok 1 and Togolok 21...All these rooms were used to make hallucinogenic drinks...in the room next to the "white room"...about ten intact as well as fragmentary earthenware stands were recovered, the last intended to support special strainers by which the juice was separated...The juice was collected in small vessels placed below the strainers, so producing a drink similar to the Zoroastrian sacred drink Haoma[.][1]

Sarianidi writes, "The Gonur temenos served the needs of the population from the nearby settlement and represented a 'sacred area' where a small temple associated with the cults of fire and hallucinogenic drinks was located behind the high and strong walls."[2] The Gonur temenos, or the "sacred area" contains the physical remains of the "cult of hallucinogenic drinks."

---

[1] Victor I. Sarianidi, "New Discoveries at Ancient Gonur," *Ancient Civilizations from Scythia to Siberia*, 2 (1995), pp. 289-293.

[2] Sarianidi, *Margiana and Protozoroastrism*, p. 119.

> In the southern, private part of the complex three rooms (178, 193 and 137) are connected with general passages to each other as well as to the public part of the complex. In the western part of room 137, there is a small mound in which three ceramic vessels have been embedded and coated with the same white gypsum plaster that covers the walls and floor. In gypsum samples taken from inside the bowls...[contained the plant] used in the preparation of the *haoma/soma* type hallucinogenic beverages used for libations. The passage from this 'white room' leads into a larger room (193) with benches built into the northern wall. On the floor of this room there were found 10 ceramic pot-stands...which seem to be associated with the preparation of the hallucinogenic drink and were perhaps used in conjunction with ceramic strainers...It is possible that the beverage was prepared in rooms 178 and 193 and then transported through the passages to the public part of the complex where ritual libations were performed in the courtyard...Although the purpose of many of the rooms remains unclear, special attention should be paid to room 330 in the corner of which is a small, gypsum-coated basin...In the same room are found ceramic stands and strainers leading to the conclusion that this may also have been a 'white room' associated with the preparation of hallucinogenic beverages.[1]

Sarianidi summarizes the archaeological evidence excavated from the temples of Margiana:

> In each of these three temples the main place is occupied by the so-called "white rooms" with a common layout principle. Along the walls of these rooms there are located low brick platforms with dug-in vessels that are fixed in the platforms and that contain thick layers of gypsum...In the Gonur temenos there was found a separate "tower complex" also related to the preparation of the cult beverage. In one room on the floor there was a large basket lined inside with a thick layer of gypsum...The archaeological excavations of the Margiana temples have yielded huge vats, "small baths" (and sometimes weaved baskets) that are plastered inside with gypsum layers...

[1] Viktor I. Sarianidi, "Temples of Bronze Age Margiana: traditions of ritual architecture," F. Hiebert (trans.), *Antiquity*, 68 (1994), pp. 388-390.

the excavations of the Gonur temenos are very significant. There, around a small temple there were scattered a lot of private houses the inhabitants of which were engaged in the everyday service of the temple. Over twenty five rooms found in these private houses have yielded either large vats or "small baths" made in the special brick platforms...Numerous stone articles connected with grinding... were found in all Margianian temples...The archeological finds show that this final stage (i.e., fermentation) of the preparation of the cult beverage took place in the above-mentioned "white rooms" of the Margiana temples since all of them along their walls had brick platforms with dug-in vessels that contained remains... After the fermentation process was finished they had to separate the intoxicating drink...and special strainers were used for this purpose. On the bottom of each strainer there was a hole covered with a piece of wool, a fact that is mentioned in detail in the Rigveda. The excavations have yielded the so-called ceramic stands found in all three temples of Margiana, as well as special strainers with centrally located holes. Of outstanding interest was a large room in the Gonur temenos that was located next to the white room. There on the floor and benches along the walls were found five intact round ceramic stands and fragments of three more, as well as large fragments of conic strainers with centrally located holes. It seems quite natural to suppose that such strainers with holes covered with pieces of sheep wool were placed on the above-mentioned "ceramic stands"...In all three Margianian temples vessels were found with long spouts as well as vessels with frail sculptural friezes along the rim. Especially the latter finds have an important meaning since their decorated rims deny their everyday usage and most likely indicate their cult purpose. The vessels with four spouts and sculptured images of goats standing by the "tree of life" were most probably connected with the cult of libation as well.[1]

[1] Victor I. Sarianidi, "Margiana and Soma-Homa."

Chapter 10: The Contents of the Temple Vessels

Beyond Wasson's *Amanita muscaria* theory,[1] and ergot-infected cereals,[2] there have been other plants proposed as candidates for Soma/Haoma.[3] Flattery and Schwartz have advanced the candidate wild rue, *Peganum harmala*,[4] as the original Soma,[5] a plant once used in Zoroastrian circles, although their hypothesis has been rejected as unconvincing.[6] Ephedra (*Ephedra distachya*)

[1] "The contention by Wasson that soma is irrefutably and without a doubt the basidiomycete *Amanita muscaria* is disconcerting." William Emboden, *Narcotic Plants*, New York, Macmillan Publishing Co., Inc., 1979, p. 61.

[2] Shelley, *Elixir*, pp. 13-29.

[3] Wendy Doniger O'Flaherty, "The Post-Vedic History of the Soma Plant," in Wasson, *Soma*, pp. 95-147.

[4] Flattery and Schwartz, *op. cit.*

[5] "[F]eatures of *Peganum harmala* which do not fit very well with the general picture of *soma/haoma* are as follows...if the use of *P. harmala* were an ancient Indo-Iranian custom, the areas where the plant commonly occurs seem to disagree with the proposed original or secondary homelands of the Indo-Aryans...the harmaline alkaloid amount is highest in the ripe seeds...but there is no textual connection (Ṛgvedic or Avestan) between *soma/haoma* and plant seeds. Instead, it is repeatedly stated in the ègveda that stems are identical to *soma*. The Vedic word usually translated as 'stem', 'stalk', 'filament', is *amśu*...but there is nothing 'filamentous' in *P. harmala*...the alkaloids of *P. harmala* have a sedative, not stimulating, effect." Harri Nyberg, "The problem of the Aryans and the Soma: The botanical evidence," *The Indo-Aryans of Ancient South Asia: Language, Material Culture and Ethnicity*, G. Erdosy (ed.), Berlin, Walter de Gruyter, 1995, pp. 389-390; "The Douvans of Bokhara used to inhale the smoke of burning *Peganum harmala* seed and became quite exuberant..." Emboden, *op. cit.*, p. 125.

[6] "The authors have conclusively shown that the rue, Peganum harmala, was used as a hallucinogenic drug in Zoroastrian circles some time before A.D. 900. The plant was given the same high respect and some of the epithets of the Haoma of old. But all attempts to connect this plant with the one in vogue more than

has also been advanced as a candidate for Soma.[1] There are approximately 40 species of Ephedra, a leafless bush growing to variable heights (0.2-4.0 m), it possesses a twisted trunk with numerous green or yellowish stems, the marrow parts of these stems are either brown-colored or colorless, and the psychoactive alkaloids L-ephedrine and pseudo-ephedrine are found mainly in the green part of the plant.[2] Some species of Ephedra, including *Ephedra distachya*, have fruit covered with an orange film.[3] To this day the Parsis use a species of this plant in their Haoma rituals.[4] Aurel Stein writes:

> Far more difficult it is to explain how the Ephedra plant came to be used for supplying the juice which in the Zoroastrian ritual practice of the present day, both among the "Gabar" communities of Yezd and Kirmān and the Pārsīs of India, figures as the representative of the ancient *Haoma*...Yet it is obviously impossible to reconcile the character of the juice obtained from this Hūm or Ephedra plant, extremely bitter and far from palatable even as a medicine, with what Rigveda hymns and the Avesta often indicate as to the exhilarating and exciting effects of both Soma and Haoma. It is clear enough that on Iranian ground, too, a substitution for the original plant must have taken place such as Sanskrit texts directly attest for

2000 years earlier are not convincing." Harry Falk, "Soma I and II," *Bulletin of the School of Oriental and African Studies*, 52 (1989), p. 77 n. 11. "[T]he identification of Haoma with *Peganum harmala*...remains very doubtful. The use of harmala is widely attested in Central Asia among Iranian peoples, but mostly for fumigation, and Haoma was pounded, not burnt." I. Steblin-Kamenskij, Review of *Papers in honour of Professor Mary Boyce*, *Bulletin of the School of Oriental and African Studies*, 50 (1987), pp. 376-378, p. 377.

[1] Falk, *op. cit.*, pp. 77-90; Nyberg, *op. cit.*, pp. 393-401.

[2] Nyberg, *op. cit.*, pp. 393-394.

[3] N. R. Meyer-Melikyan, "Analysis of Floral Remains from Togolok-21," in Sarianidi, *Margiana and Protozoroastrism*, Appendix II, p. 178.

[4] "[T]he pharmacological action of ephedrine as a stimulant is of insufficient intensity and too inconsistent in character with what is indicated for sauma by the Iranian evidence to allow identification of *Ephedra* itself as sauma...The clearest demonstration that *Ephedra* cannot have been sauma exists in the very fact that *Ephedra* extracts are today drunk as *haoma* by Zoroastrian priests who do not become intoxicated from them and whose predecessors, moreover, apparently consumed *Ephedra* extracts for centuries without so much as knowing that *haoma* is said in the Avesta to be intoxicating...*Ephedra* is without suitable psychoactive potential in fact...and, therefore, it cannot have been believed to be the means to an experience from which priests could claim religious authority or widely believed to be the essential ingredient of an *intoxicating* extract." Flattery and Schwartz, *op. cit.*, pp. 72-73.

India in the case of the original Soma of the Vedic hymns.[1]

Bone tubes made from the midsections of sheep-goat femora (9 to 11 cm) that were cut, polished, reamed out, and incised were recovered from sites in Margiana and Bactria. Sarianidi has interpreted these as sipping tubes for an opium-ephedra concoction, a conjecture with several problems. Opium is obtained from *Papaver somniferum* L., which is a hybrid species that does not occur in the wild, and requires cultivation.[2] There currently exists no evidence that *Papaver somniferum* was cultivated at Margiana, or that the opium poppy was known in Central Asia in the second millennium B.C. There is no mention of the poppy in cuneiform Sumerian, Akkadian, or Assyrian writings.[3] Furthermore, the texts show that *Papaver somniferum* was unknown in India before the seventh century A.D., the poppy appearing as a medicine in the Ayurvedic treatises of the eighth century.[4] Hiebert writes:

> These pieces have been found so far only in architectural context, not in burials. Opium poppy pollen has been recorded from soil inside of the tubes leading Sarianidi to interpret them as drinking tubes for sipping an opium-ephedra drink similar to *haoma*. It must be noted, however, that poppy flowers are a common spring annual in the Kara Kum desert, and that pollen studies have not been conducted from other soils in the Bronze Age sites in Margiana. It is likely that poppy was part of the natural background in the pollen profile of the ancient oases and that the function of the tubes was unrelated to the pollen.[5]

It is the opinion of Sarianidi that the basin in room 330 of the Gonur

---

[1] Aurel Stein, "On the Ephedra, the Hūm Plant, and the Soma," *Bulletin of the School of Oriental and African Studies*, 6 (1931), p. 505.

[2] Péter Tétényi, "Opium Poppy (*Papaver somniferum*): Botany and Horticulture," *Horticultural Reviews*, 19 (1997), pp. 373-408.

[3] "Dr. Hans Helbaek, Danish archeologist and recognized world authority on plant collecting and prehistoric agriculture in the Near East, stated in a letter to me dated April 24, 1973, that 'among all the gallons of grain from Persia, Iraq, Turkey, Lebanon, Syria, Jordan and Egypt I have never established *P. somniferum*. These finds reach in time from c. 7500 B. C. to A. D. 1000.'" Abraham D. Krikorian, "Were the Opium Poppy and Opium Known in the Ancient Near East?," *Journal of the History of Biology*, 8 (1975), pp. 95-114; Tétényi, *op. cit.*, p. 400.

[4] P. G. Kritikos and S. P. Papadaki, "The history of the poppy and of opium and their expansion in antiquity in the eastern Mediterranean area," G. Michalopoulos (trans.), *Bulletin on Narcotics*, 19 (1967), p. 38.

[5] Hiebert, *Oasis Civilization*, p. 146.

temenos contains the remains of hemp (*Cannabis sativa*),[1] and in the opinion of Meyer-Melikyan the numerous impressions made by the organic remains that were once in the vessels in room 137 were from hemp seeds.[2] According to Meyer-Melikyan and Avetov, the gypsum samples from the ceramic vessel from the Gonur temenos tested positive for Ephedra, poppy, and hemp.[3] Meyer-Melikyan also analyzed samples from Togolok-21, including a bone tube, and reported the presence of Ephedra and poppy.[4] Harri Nyberg at the Department of Botany, University of Helsinki, also tested samples from these sites:

> The earliest archaeological records about the use of ephedras come from Turfan...where Sir Aurel Stein (Stein 1931) found bunches of broken plant twigs in several graves...his floral samples were examined at Kew Gardens in 1984 and were found to be the remains of horsetails (*Equisetum L., Equisetaceae*)...remains of ephedras have also been reported from the temple-fortress complex of Togolok 21 in the Merv oasis...along with the remains of poppies (*Papaver L.*, but is unclear whether the species was *P. somniferum*, *P. setigerum* or some other poppy). In 1991, I received some samples from the site, which were subjected to pollen analysis...in most cases only pollen of the family Caryophyllaceae was found, along with some pollen remains from the families Chenopodiaceae and Poaceae (grain crops?). The largest amou[n]t of pollen was found in a bone tube (used for imbibing liquid?) from Gonur I, but even in this sample...only pollen of the family Caryophyllaceae was present. No pollen from ephedras or poppies was found...[5]

[1] Although the cannabis plant was utilized by particular Scythian tribes in the first millennium B.C., it was not pounded and swallowed as a fermented liquid, but was burnt and inhaled. Nor did these tribes use cannabis in their religious rituals. Herodotus writes, "The Scythians then take the seed of this hemp and, creeping under the rugs, they throw it on the red hot stones; and, being so thrown, it smoulders and sends forth so much steam that no Greek vapour-bath could surpass it. The Scythians howl in joy for the vapour-bath. This serves them instead of bathing, for scarce ever do they wash their bodies with water." Herodotus, *Histories* IV.75.

[2] Sarianidi, *Margiana and Protozoroastrism*, p. 113.

[3] N. R. Meyer-Melikyan, and N. A. Avetov, "Analysis of Floral Remains in the Ceramic Vessel from the Gonur Temenos," in Sarianidi, *Margiana and Protozoroastrism*, Appendix I, pp. 176-177.

[4] Meyer-Melikyan, *op. cit.*, pp. 178-179.

[5] Nyberg, *op. cit.*, pp. 399-400.

As one would expect, several forms of barley and wheat have been identified from the Gonur samples.[6] C. C. Bakels and others have determined that the archaeological evidence contradicts Sarianidi's multi-plant Soma/Haoma theory (Ephedra, opium, and hemp), and that the gypsum deposits in the Margianian vessels contain the remains of broomcorn millet (*Panicum miliaceum* L.). It is of interest that broomcorn millet has not been identified from Gonur-depe samples, which suggests the possibility that the cereal was grown elsewhere and transported to the temple for ritual purposes.[7] *Panicum miliaceum* ranks among the hardiest cereals, growing in poor soils it can stand up well against intense heat and severe drought,[8] and is a host to *Claviceps purpurea*.[9] Bakels writes:

> Vessels found in the "white room" of the Gonur temenos and in Togolok-21 revealed part of their original contents as holes in a gypsum and clay deposit on their bottom. The holes are the negatives of plant matter which itself has decayed. The white layer of gypsum and clay has been separated from the ceramic fabric of the vessels and parts of it have reached my laboratory for an identification of the plants, which have left their imprints...Most of the impressions are round to oval. A small minority has clearly been left by stems. The round impressions have been published as having been left by hemp seeds (Cannabis sativa) and the stems by Ephedra. The material sent to me reveals, however, neither of these. The impressions caused by seeds are not of hemp. They are too small, for instance, do not have the right shape nor the right type of surface pattern. The long, grooved stems are not incontestably identifiable as Ephedra. The original contents consisted in my opinion of broomcorn millet (Panicum miliaceum) and the stems might also belong to this cereal, although that cannot be proven. Some of the round impressions still contain a cell layer resembling a cell layer of broomcorn millet husks. They are preserved because of their high silica content. My interpretation is that the vessels were filled with

6 Hiebert, *Oasis Civilization*, p. 132.

7 Mark Nesbitt, "Plant Use in the Merv Oasis," *Iran: Journal of the British Institute of Persian Studies*, 35 (1997), pp. 29-31.

8 Zohary and Hopf, *op. cit.*, pp. 83-86.

9 Bove, *op. cit.*, p. 47.

not yet dehusked broomcorn millet. To obtain a second opinion I showed the material to Sietse Bottema and René Cappers from the University of Groningen in the Netherlands. They had in their reference collection small-seeded hemp from Iran, but these were still too large, and again, the overall form and the surface pattern did not fit. Both colleagues were of the opinion that the impressions were left by a millet, presumably broomcorn millet. In addition I had the opportunity to show the material to Mark Nesbitt from the Centre for Economic Botany, Royal Botanic Gardens in Kew, Great Britain, who is familiar with material from the Merv oasis and to Dorian Fuller from the Institute of Archaeology, London, Great Britain, who is an expert on Asian millets. Both colleagues came to the conclusion that broomcorn millet provides the best fit. The original publication mentions also pollen, hemp pollen grains in large quantities, but also pollen from other plants. I did not succeed in extracting pollen from the white substance…The material we examined contained broomcorn millet. This cereal is known from the Merv oasis, at least from the Bronze Age onwards. The crop plant most probably has its origin in Central Asia, perhaps even in the Aralo-Caspian basin. It is a cereal that can be cooked, made into a heavy bread, or used to prepare a fermented drink. The latter can be done with undehusked grain.[1]

[1] C. C. Bakels, "The contents of ceramic vessels in the Bactria-Margiana Archaeological Complex, Turkmenistan," Electronic Journal of Vedic Studies, 9 (2003).

Chapter 11: The Mysteries of Eleusis

The Eleusinian Mysteries were the most important religious rites performed in ancient Greek Civilization. With the exception of one year, the initiation into the Greater Mystery took place in the fall every year for nearly twenty centuries, performed in the Telesterion at Eleusis, in Attica, near Athens. The rituals began in the spring with the Lesser Mystery at Agrae, which is also near Athens. The ritual of the Greater Mystery centered around the legend of the Goddess Demeter, and her daughter Persephone, who was also called Kore (Greek "The Maiden"). The Greeks called ergot-infected cereal *erysibe*, or "rust," and this was also an epithet of Demeter, as well as the Sun-god Apollo.[1] The Egyptian Goddess Isis was credited with the discovery of health-giving drugs, and was honored with Demeter as the inventor of grain (*frumentum-frugmentum*), fruit (*fruges*), and the legal order.[2]

The myth connected with the Mystery at Eleusis is contained in the *Homeric Hymn to Demeter*, in which Hades kidnaps the young Goddess Persephone while she is picking flowers, causing her mother Demeter to become so distraught by her loss that she causes a drought that threatens to destroy the human race. This alarms the gods, who then persuades Hades to release his bride, which he does, but not before giving her a "pomegranate," and when Persephone returns, she discovers that she has consumed the fruit

[1] *Etymologicum Gudianum* CCX.25; Strabo, *Geographica* XIII.1.64.
[2] Diodorus Siculus, *The Library of History* I.25.2; Lucretius, *De Rerum Natura* II.549-560; etc.

of the underworld, and now must return there every year for a period of time.

The abduction of Persephone represents the divine possession of the frenzy-god Dionysus, who is Hades in this context.[1] Dionysus, the "Divine Bridegroom," was identical to his father Zeus, Hermes, Iakchos, and was also called Iasion.[2] According to legend, Demeter became intoxicated on the divine nectar and copulated with Iasion on the Rarian Plain, conceiving Plutus.[3] Plutus was the god of wealth and riches, his name also being an epithet for Hades.[4] In the Greater Mystery the psychopomp Hermes Triptolemus led the initiates to their salvation with his sacred plant called moly. Plato seems to look upon Hermes and the Logos, the Word of God, as being one and the same.[5] Philostratus calls moly the symbol of "fellowship with the Logos and of spiritual strivings."[6] Eustathius writes, "For moly comes from God and is a gracious gift."[7] Triptolemus was held to be the first to have sown grain for cultivation, traveling the world in a serpent-drawn chariot.[8] Triptolemus was said to have been taught the mysteries of mortality and rebirth by Demeter, and was held in non-Orphic tradition to be the son of Raros, whose name relates to the Rarian plain, where the sacred plant was gathered for the secret Mystery rite.[9]

In the secret rites of the Eleusinian Mysteries, the initiates of the Greater Mystery consumed a visionary preparation known as the *kykeon*, or mixed drink, and in the Lesser Mystery, a Dionysian rite that was later incorporated into the institution, the participants also ingested a secret potion. The Hierophant priests belonging to the Athenian family of the Eumolpidai, named after the Thracian king Eumolpus, conducted these rites, which were performed for nearly two thousand years until a Mithraic priest usurped the Hierophant's throne; and fulfilling the prophecy of the rival he replaced, the Visigothic king Alaric I, in the company of his Christian monks in black robes, invaded Greece and destroyed the sanctuary in A.D. 396, and with this

[1] Heraclitus, *Fragment* 15.
[2] Herodotus, *Histories* VIII.65.
[3] Homer, *Odyssey* V.125-128; Hesiod, *Theogony* 969-974.
[4] Euripides, *Alcestis* 360.
[5] Plato, *Phaedrus* 264C; Plato, *Cratylus* 407E.
[6] Philostratus, *On Heroes* 665.
[7] Eustathius, *Odysseum* 1658.
[8] Pausanias, *Description of Greece* I.14.3.
[9] Ibid. I.38.6.

came the end of Greek civilization.[1]

Heracles was among those initiated into the Greater Mystery at Eleusis, and prior to that he claimed that he had no need for initiation because he had already had a vision of the Goddess when he descended into Hades.[2] Apollodorus also records the vision of Persephone by Heracles.[3] According to Euripides, the vision Heracles received at Eleusis allowed him to conquer death and return to the other world like the Goddess.[4]

According to Aristotle, the Mystery at Eleusis was something experienced rather than something learned.[5] Iamblichus writes, "It is necessary, therefore, from the beginning, to divide ecstasy into two species, one which leads to a worse condition of being and fills us with stupidity and folly, but the other imparts goods which are more honorable than human temperance."[6]

In among other sources, the testimony contained in the *Homeric Hymn to Demeter* reveals the significance of the Greater Mystery witnessed at Eleusis: "Whoever among men who walk the earth has seen these Mysteries is blessed, but whoever is uninitiated and has not received his share of the rite, he will not have the same lot as the others, once he is dead and dwells in the mould where the sun goes down."[7] In a poem Sappho proclaims: "For death is all I wish for me, and the dew lotus fields to see, the meadows of Elysium."[8]

Pausanias writes, "For the Greeks of an earlier period looked upon the Eleusinian mysteries as being much higher than all other religious acts as gods are higher than heroes."[9] Isocrates reminded the Athenians that Demeter granted agriculture by which "man rises above the beasts," and initiation, which brings hope regarding "the end of life and all eternity."[10] According to Cicero, the initiates at Eleusis obtained a serener life together with a hope of a better death, and writes: "among the excellent and divine gifts to Athens, to

[1] C. Kerényi, *Eleusis: Archetypal Image of Mother and Daughter*, R. Manheim (trans.), New York, Bollingen Foundation, 1967, pp. 16-18.
[2] Xenophon, *Hellenica* VI.3.6; *Papiri della Reale Universita di Milano*, I (ed. Voliano), Milan, 1937, p. 177.
[3] Apollodorus, *The Library* II.5.12.
[4] Euripides, *Madness of Heracles* 613.
[5] Aristotle, *Fragment* 15; in Synesius, *Oration* 48.
[6] Iamblichus, *Mysteries of the Egyptians, Chaldeans, and Assyrians* III.25.
[7] *Homeric Hymn to Demeter* 479-483.
[8] Sappho, *Berlin Fragment* 9722.4.
[9] Pausanias, *Description of Greece* X.31.11.
[10] Isocrates, *Panathenaicus* 28.

the life of men, nothing is better than those Mysteries by which we are drawn from savagery to civilization."[1]

[1] Cicero, *De Legibus* II.14; II.36.

Chapter 12: Proposed Methods of Preparing the *Kykeon*

In the following pages, the existing proposals regarding the ergot theory and the method of preparing the *kykeon* are examined, and a new proposal is advanced: the acid-catalyzed isomerization of the alkaloids in ergot, followed by the alcohol extraction of isolysergic acid hydroxyethylamide, the isomer of lysergic acid hydroxyethylamide. This process simply involves the boiling of ergot for a number of hours in an acidic solution, and fermentation. Lysergic acid hydroxyethylamide occurs in *Claviceps purpurea* (Fr.) Tul., which is parasitic to cultivated cereals and wild grasses. *Claviceps purpurea* regularly infects the wild grasses of *Lolium* spp., including *Lolium temulentum* L., also known as darnel, cockle, or tares, a common weed among cereal crops. In his work entitled *A Dictionary of Assyrian Botany*, R. Campbell Thompson offers compelling evidence that ergot-infected darnel (*L. temulentum*) was employed as a medicine in the Old World, often in beer, and the drug was also macerated in wine to produce intoxication.[1] The toxic peptide alkaloids in ergot are insoluble in water, but these are all to some extent soluble in alcohol.[2] This begs the question: how did the Assyrians accomplish this

[1] R. Campbell Thompson, *A Dictionary of Assyrian Botany*, London, British Academy, 1949, pp. 56, 146-148.

[2] *CRC Atlas of Spectral Data and Physical Constants for Organic Compounds*, Second Edition, J. G. Grasselli and W. M. Ritchey (eds.), Cleveland, CRC Press, Inc., 1975, vol. III, pp. 228-229.

without serious, if not fatal consequences?[1]

In *The Road to Eleusis*, Albert Hofmann proposed the water-soluble theory as the method of preparing the *kykeon*, specifically from *Claviceps purpurea* infecting barley and *Lolium temulentum*, and from *Claviceps paspali* Stevens and Hall infecting the grass *Paspalum distichum* L.[2] Hofmann's method involved the "separation of the hallucinogenic agents by simple water solution from the non-soluble ergotamine and ergotoxine alkaloids."[3] Hofmann writes: "These alkaloids, mainly lysergic acid amide (ergine), lysergic acid hydroxyethylamide, and ergonovine, are soluble in water, in contrast to the non-hallucinogenic medicinally useful alkaloids of the ergotamine and ergotoxine type."[4]

However, ergonovine (ergometrine, lysergic acid propanolamide) is a powerful uterine stimulant that has been connected with gangrene and other unpleasant symptoms, and this alkaloid is not hallucinogenic.[5] Along with Richard Evans Schultes, Hofmann states this in *The Botany and Chemistry of Hallucinogens*: "Psychotomimetic effects are unknown for ergometrine, which is used to a large extent in obstetrics as a uterotonic and haemostatic agent. In small dosages administered for this purpose, the alkaloid apparently has no action on psychic functions. Its occurrence in the alkaloid mixture of ololiuqui can thus have no significant effect on the mental action."[6] The visionary ololiuqui of the Aztecs has been conclusively identified as two species of morning glory seeds (*Rivea corymbosa* (L.) Hallier f. and *Ipomoea violacea* L.), a topic that will be revisited in later pages. Thomas J. Riedlinger

[1] "*L. temulentum*, or darnel...is remarkable as being the only well authenticated instance of a plant belonging to the order of Grasses, in which narcotic or even deleterious properties have been found. The grains are said to produce intoxication in man, beasts, and birds, and to bring on fatal convulsions. According to Christison, darnel, when mixed with flour and made into bread, has been known to produce headache, giddiness, somnolency, delirium, convulsions, paralysis, and even death." *The Penny Cyclopædia of the Society for the Diffusion of Useful Knowledge*, London, Charles Knight and Co., 1839, vol. XIV, p. 99.

[2] Hofmann, *Road to Eleusis*, pp. 33-34.

[3] Ibid., p. 33.

[4] Ibid., p. 32.

[5] George Barger, "The Alkaloids of Ergot," *Handbuch der Experimentellen Pharmakologie*, W. Heubner and J. Schüller (eds.), Berlin, Julius Springer, 1938, vol. VI, pp. 179-182, 197; M. E. Jarvik *et al.*, "Comparative Subjective Effects of Seven Drugs including Lysergic Acid Diethylamide (LSD-25)," *Journal of Abnormal and Social Psychology*, 51 (1955), pp. 657-662.

[6] Schultes and Hofmann, *op. cit.*, p. 154.

summarizes the material related to the ergonovine hypothesis:

> Hofmann (Wasson et al. 1978:30-31) did report "psychotropic, mood-changing, slightly hallucinogenic activity" from a self-administered 1.5 mg dose of ergobasin (the ergonovine base of a 2.0 mg dose of ergonovine hydrogenmaleinate)...But Wasson (1976) had earlier reported, in a letter to Hofmann dated 22 May 1976, that when he and a colleague ingested almost identical doses (adjusted for body weight) they were "strongly affected psychically but were without hallucinations, visual or auditory..." Results of later tests by other researchers who ingested substantially higher, 10.0 mg doses of ergonovine (Bigwood, Ott & Neely 1979) and 2.0 mg doses of the more potent lysergic acid derivative methylergonovine (Ott & Neely 1980) were likewise disappointing. The latter concluded that both drugs, unlike LSD, evoke uncomfortable somatic effects, such as lassitude, that "supervene at doses insufficient to elicit profound entheogenic (hallucinogenic) effects."[1]

Following the objections raised against Hofmann's water-soluble hypothesis, Peter Webster and Daniel M. Perrine proposed the method of "partial hydrolysis." In the procedure described by the authors, *Claviceps purpurea* is boiled in an alkaline solution produced by wood ash (with the addition of ethyl alcohol), converting lysergic acid hydroxyethylamide and the toxic peptide alkaloids to ergine (lysergic acid amide) and its isomer isoergine (isolysergic acid amide). Referring to the process employed by Smith and Timmis with which they obtained ergine from ergotinine,[2] which consists of ergocornine, ergocristinine, and ergokryptinine, Perrine writes: "*We suggest that conditions of solubility, pH, and temperature which would be equivalent in effect to ST's process, and result in the conversion of ergot to ergine, could be readily obtained by boiling crude ergot for several hours in water to which the ashes of wood or other plant material, perhaps barley, had been added.*"[3] Perrine adds: "The only other likely psychoactive ingredient in

[1] Riedlinger, *op. cit.*, p. 155.

[2] Sydney Smith and Geoffrey Millward Timmis, "The Alkaloids of Ergot. Part III. Ergine, a New Base obtained by the Degradation of Ergotoxine and Ergotinine," *Journal of the Chemical Society* [London], 1932, pp. 763-766.

[3] Peter Webster, Daniel M. Perrine, and Carl A. P. Ruck, "Mixing the *Kykeon*," *Eleusis*, New Series, 4 (2000), p. 71.

the *kykeon* would be wine."[1] Webster proposed a short period of boiling under the same conditions:

> The details of the various experiments indicated that longer reaction times and higher temperatures favoured the complete transformation to lysergic acid, while short reaction times and lower temperatures resulted in significant amounts of ergine. Subsequent unpublished work by myself indicated the strong dependence of the hydrolysis on temperature, but weak dependence on base concentration…Is it possible that merely digesting powdered ergot with wood ash and water (and possibly wine containing 10% or so of ethanol to improve the solubility of the alkaloids), heating the mixture for a short period and then filtering off the liquid might have been the method of mixing the *kykeon?*[2]

There is one insurmountable problem with this proposal: ergine and isoergine possess no hallucinogenic activity. Webster writes: "Although several self-experiments with *ololiuhqui* (seeds of two species of morning glory) or with its purified alkaloids ergine and isoergine have been inconclusive…Ergine and its stereoisomer isoergine are the two principal psychoactive compounds of *ololiuhqui*."[3] Perrine includes the following evidence suggesting that ergine (lysergic acid amide) and isoergine are not hallucinogenic, including the material related to the visionary ololiuqui of the Aztecs:

> 1 – In 1955, Humphrey Osmond took 60 to 100 seeds and experienced, according to Schultes and Hofmann, "a state of apathy and listlessness, accompanied by increased visual sensitivity. After about four hours, there followed a period in which he had a relaxed feeling of well-being that lasted for a longer time."
>
> 2 – Hofmann compares the effects of *ololiuhqui* rather unfavorably with those of LSD in his book, *LSD: My Problem Child.* His description is as follows: "*After the discovery of the psychic effects of LSD, I had also tested lysergic acid amide [lysergamide] in a self-experiment and established that it likewise evoked a*

[1] Ibid., p. 65.
[2] Ibid., pp. 62-63.
[3] Ibid., p. 59.

dreamlike condition, but only with about a tenfold to twentyfold greater dose than LSD. This effect was characterized by a sensation of mental emptiness and the unreality and meaninglessness of the outer world, by enhanced sensitivity of hearing, and by a not unpleasant physical lassitude, which ultimately led to sleep." As for the effects of the Mexican morning-glory seeds, he continues: "*The psychic effects of ololiuhqui, in fact, differ from those of LSD in that the euphoric and the hallucinogenic components are less pronounced, while a sensation of mental emptiness, often anxiety and depression, predominates. [Such]...weariness and lassitude are hardly desirable effects...in an inebriant.*"

3 – Solms, in a systematic comparative study of the psychopharmacology of lysergamide, confirmed Hofmann's impressions, concluding that it "induces indifference, a decrease in psychomotor activity, the feeling of sinking into nothingness and a desire to sleep… until finally an increased clouding of consciousness does produce sleep."

4 – Hofmann also tested the effects of isoergine…and found that 2.0 mg orally produced a syndrome not very different from ergine, characterized by sensations of "tiredness, apathy, a feeling of mental emptiness and the unreality and complete meaninglessness of the outside world."[1]

To counter this evidence for the absence of hallucinogenic activity in ergine and isoergine, Perrine compares the effects of these alkaloids to *Amanita muscaria*, which is also not hallucinogenic:

And the principle psychoactive agent in *ololiuhqui* (*Rivea corymbosa, Ipomoea violacea*) is just what we propose to have been the active agent in the *kykeon*: ergine (with smaller amounts of isoergine)…Indeed, other *pharmaka* with far less likely a priori plausibility if judged solely by their clinical psychopharmacology have been exploited for millennia in various cultures for their entheogenic effects: I am thinking particularly of *Amanita muscaria* (which despite its likely identity with the archetypical Vedic entheogen, *soma,* produces in secular settings only a bizarre state of

[1] Ibid., p. 65.

incoherent delirium)...[1]

Perrine writes: "No one disputes that ergine unmistakably and profoundly alters everyday consciousness — the only dispute concerns whether this alteration has, considered in isolation, properly 'entheogenic' qualities."[2] Albert Hofmann reported the following effects of ergine in a self-experiment conducted on 30 October 1947:[3]

> 10:00h: Intramuscular injection of 0.5 ml of 1 per mille solution of LA 111 (=0.5 mg d-lysergic acid amide).
>
> 11.00h: Tiredness in the neck, slight nausea.
>
> 11.05h: Tired, dreamy, incapable of clear thoughts. Very sensitive to noises which give an unpleasant sensation.
>
> 11.10h: Desire to lie down and sleep. Genuine physical and mental tiredness, which is not experienced as an unpleasant sensation. Slept for 3 hours.

---

[1] Perrine, "Mixing the *Kykeon*," p. 66.

[2] Ibid., p. 68.

[3] The year that R. Gordon Wasson, Albert Hofmann, and Carl A. P. Ruck published *The Road to Eleusis* (1978) containing Wasson's ergot theory, and Calvert Watkins published "Let Us Now Praise Famous Grains," which is cited in the work, Wasson would write to Hofmann: "I foresee that our *Eleusis* may lead to a challenge of my identity for Soma in *SOMA: Divine Mushroom of Immortality*." With the knowledge that he had potentially created two mutually exclusive theories, the following year Wasson and others proposed in the *Journal of Psychedelic Drugs* that the words hallucinogen and psychedelic should be replaced by the term "entheogen," defined by the authors as "becoming the gods within." The following year, at a time when the so-called "counter-culture" generation of the 1960s was basically irrelevant, Wasson contended that terms such as hallucinogen and psychedelic had become vulgarized "by hippie abuse," and repeated this charge six years later when he wrote that these terms possess the "odor...of the youth of the 1960s," an ironic statement considering the usage of the term entheogen today. Wasson also defined entheogens as "substances that keep one, willy-nilly, wide awake," a definition that certainly does not square with the effects of *A. muscaria*. This deliberate maneuver in semantics allowed Wasson to reclassify *A. muscaria*, a deliriant, with a specific class of plants and drugs, including ergot and certain amide derivatives of lysergic acid, that are scientifically classified as hallucinogenic. New terms in science are created to be more specific, not more vague. As such the term "entheogen" is completely devoid of any scientific reason for existing and it deliberately confuses established scientific terminology. Calvert Watkins, "Praise Famous Grains," pp. 9-17; Wasson to Hofmann, 19 July 1978; in Riedlinger, *op. cit.*, p. 154; Carl A. P. Ruck, Jeremy Bigwood, Danny Staples, Jonathan Ott, and R. Gordon Wasson, "Entheogens," *Journal of Psychedelic Drugs*, 11 (1979), pp. 145-146; R. Gordon Wasson, *The Wondrous Mushroom: Mycolatry in Mesoamerica*, New York, McGraw-Hill, 1980, pp. xiv, xxiv; R. Gordon Wasson, Stella Kramrisch, Jonathan Ott, and Carl A. P. Ruck, *Persephone's Quest: Entheogens and the Origins of Religion*, New Haven, Yale University Press, 1986, p. 124.

15.00h: Return of normal condition with full capacity for performing work.[1]

According to Fanchamps, hallucinogenic activity is completely absent from ergine: "*d-lysergic acid amide* (LA 111, ergine, No. 18) is not hallucinogenic; in doses up to 1 mg i.v., it produces — besides autonomic disturbances such as hypersalivation, emesis, dizziness and diarrhea — sedation, clouding of consciousness, and finally sleep (SOLMS, 1956a, 1956b; ISBELL, 1962)."[2] On the subject of ololiuqui, Fanchamps writes:

> Ololiuqui, one of the three "magic" plants of the Aztecs, has been identified as corresponding to two varieties of morning glory, *Rivea corymbosa* and *Ipomea violacea*, the seeds of which are still being ingested by some Mexican Indians in religious or therapeutic rituals. In a series of self-experiments, OSMOND (1955) developed apathy and hypnagogic phenomena after the ingestion of seeds of *Rivea corymbosa*.
>
> Having succeeded in identifying the active principles of Ololiuqui, HOFMANN and his colleagues (HOFMANN and TSCHERTER, 1960; [etc.]) were surprised to find out that they consisted of six already-known alkaloids of the ergot group, which had all been either prepared synthetically or extracted from various varieties of *Claviceps* but never from higher plants.
>
> One of them was *d-lysergic acid amide* or LA 111 (No. 18), the autonomic and weak psychotomimetic effects of which — chiefly sedation and reduced consciousness — had already been described (p. 591).
>
> The second main component was *d-isolysergic acid amide*, or Iso-LA 819 (No. 18a). After its identification in ololiuqui, trials were performed in man with oral doses up to 5 mg; they revealed central effects which were not LSD-like (ISBELL, 1962) but chiefly consisted in relaxation, synesthesias, and altered time experience (HEIMANN, 1965; HEIM et al., 1968).
>
> *Elymoclavine* (No. 6) produces mainly sedation (ISBELL and

[1] Albert Hofmann, "The Active Principles of the Seeds of *Rivea corymbosa* and *Ipomoea violacea*," *Botanical Museum Leaflets, Harvard University*, 20 (1963), p. 209.
[2] Fanchamps, *op. cit.*, p. 591.

GORODETZKY, 1966).

Lysergol (No. 79a) has no effect up to 6 mg, but 8 mg produce a slight sedation (HEIM et al., 1968).

Ergometrine (No. 19) is a specific uterotonic and has very little central effects (JARVIK et al., 1955a).

The last component, *Chanoclavine*, is a tricyclic alkaloid, which is devoid of ergot-like activities.[1]

Recounting the history and the rediscovery of the hallucinogenic ololiuqui of the Aztecs, Schultes and Hofmann write:

> Numerous early chroniclers of the time of the Spanish conquest of Mexico reported the religious and medicinal use of a small lentil-like, hallucinogenic seed called *ololiuqui*. The Aztecs and other Indians ingested ololiuqui for purposes of divination…An early record, written in 1629, reported that "…when drunk, the seeds deprives of his senses him who has taken it, for it is very powerful." Still another source said that "it deprives those who use it of their reason. The natives…communicate with the devil…when they become intoxicated with ololiuqui, and they are deceived by the various hallucinations which they attribute to the deity which they say resides in the seeds"…One report asserted the Aztec priests… "put to it a certain seed…called *ololuchqui*, whereof the Indians make a drink to see visions"…Hernández, whose report seems to be the most reliable of the early writers, mentioned its presumed pain-dulling effects and, after a detailed description of its many medicinal uses, stated that "when the priests wanted to commune with their gods and to receive a message from them, they ate the plant to induce a delirium, during which a thousand visions and satanic hallucinations appeared to them." For nearly four centuries no convolvulaceous plant was encountered in use in Mexico as a divinatory or ritualistic hallucinogen…It was not, however, until 1939 that Schultes and Reko collected identifiable botanical material of *R. corymbosa* from a cultivated plant employed in divination by a Zapotec witch doctor in northeastern Oaxaca.[2]

[1] Ibid., pp. 594-595.
[2] Schultes and Hofmann, *op. cit.*, pp. 144-147.

Schultes and Hofmann cite the self-experiments by Osmond (1955), and the experiments of Kinross-Wright (1959), who found no ascertainable effects in doses of up to 125 seeds, and the authors report: "Alcoholic extracts produced a kind of narcosis or partial narcosis in frogs and mice."[1] The authors also state that the psychotomimetic effects of ololiuqui are produced by ergine, lysergic acid hydroxyethylamide, elymoclavine, lysergol, and possibly isoergine.[2] The effects of ergine, isoergine, elymoclavine, and lysergol have been described, and none are hallucinogenic.[3] Fanchamps writes:

> In a cross-over study on six former opiate addicts, ISBELL and GORODETZKY (1966) compared the effects of a crude extract containing the total alkaloids of *Ipomea violacea* (5 mg) to the effects of 5 mg of a synthetic mixture of the six components (LA 111 [ergine] 45%, Iso-LA 819 [isoergine] 25%, elymoclavine 5%, lysergol 5%, ergometrine 10%, and chanoclavine 10%), of 1.5 µg/kg LSD and of a placebo. This study confirmed that the crude extract and the synthetic mixture had practically identical, predominantly sedative properties and produces only slight autonomic changes; this contrasted sharply with the spectacular psychotomimetic and autonomic actions of LSD in the same subjects. In another group of addicts, 6 g of ground seeds of *Ipomea violacea* produced only very little effects.
>
> A similar study was performed by HEIMANN and his colleagues (HEIMANN, 1965; HEIM et al., 1968), comparing the artificial mixture of the six alkaloids with LA 111, Iso-LA 819, and lysergol. They found that low doses of the mixture (2-3 mg) produces a relaxation resembling the effect of Iso-LA 819, whereas high doses (6-7 mg) elicited unpleasant autonomic changes and a reduced consciousness, such as observed after LA 111.[4]

Although it does appear from the evidence that the Aztecs consumed ololiuqui in both solid and liquid form, their method of preparing these seeds has yet to be conclusively determined. Since the amide derivatives

[1] Ibid., p. 147.
[2] Ibid., p. 154.
[3] Fanchamps, *op. cit.*, pp. 591, 594-596.
[4] Ibid., p. 595.

of lysergic acid in ergot are also contained in the seeds of *Rivea corymbosa* and *Ipomoea violacea*, it seems likely that the Aztecs obtained from these seeds an hallucinogenic agent that was also present in the *kykeon*. Schultes and Hofmann concluded that ergine is the main constituent of ololiuqui,[1] and a few years later in *The Road to Eleusis*, Hofmann stated that the main constituents of ololiuqui are ergine and lysergic acid hydroxyethylamide.[2] Hofmann has subsequently written: "LSD (chemically *Lysergsäure diäthylamid* or lysergic acid diethylamide), which is closely related to lysergic acid hydroxyethylamide, the active agent of the ancient Indian 'magic drug' *ololiuhqui*."[3] The seeds of numerous species of the morning glory family (*Convolvulaceae*)[4] contain lysergic acid hydroxyethylamide, and these include *Rivea corymbosa, Ipomoea violacea, Stictocardia tiliifolia,*

[1] Schultes and Hofmann, *op. cit.*, p. 151.

[2] Hofmann, *Road to Eleusis*, p. 31.

[3] Albert Hofmann, Foreword to Myron J. Stolaroff, *The Secret Chief*, Charlotte, N.C., Multidisciplinary Association for Psychedelic Studies, 1997, p. 22.

[4] "Cohen has also reported one case of an intense psychotic reaction following the ingestion of 300 Heavenly Blue seeds, which was probably responsible for the subsequent suicide of the person in question 3 weeks later (Cohen, 1964). Long term effects resulting in a recurrence of the psychotic symptoms is known to occasionally result from the use of certain hallucinogenic drugs including LSD (Cohen, 1960) and this author points out that similar dangers could exist in cases of morning glory intoxication (Cohen, 1964). Ingram has described a psychotic episode experienced by a 20-year old university student, following the ingestion of 250 *Ipomoea* seeds, which was sufficiently severe to require hospitalisation (Ingram, 1964). This author also underlines the danger that latent psychoses may be activated by excessive ingestion of such seeds (Ingram, 1964). In 1966, a study of three cases of morning glory seed intoxication showed that the effects observed were similar to those following LSD ingestion (Fink *et al.*, 1966). The possibility of additional toxic reactions resulting from the ingestion of such seeds, which have been treated with potentially toxic fungicides and pesticides to prevent spoilage on storage, should also not be overlooked...(Savage and Stolaroff, 1965) report that low doses (20-50 seeds) of Heavenly Blue seeds are capable of inducing 'beginning imagery' and that higher doses (100-150 seeds) produce definite psychological effects (i.e. spatial distortions, visual and auditory hallucinations and other effects characteristic of the LSD experience)...It has also been reported that extracts of seeds of the psychotomimetic *I. violacea* morning glory varieties will produce a definite uterine stimulant effect due to the presence in the extracts of ergometrine, a compound with known oxytocic properties (Der Marderosian *et al.*, 1964). It is estimated that 500 Heavenly Blue or Pearly Gates seeds could contain up to 1 mg of ergometrine and it is usually considered that between 0.2 and 0.5 mg of (ergometrine) is oxytocic in humans. Several workers have warned against the dangers of ergot poisoning due to excessive morning glory seed ingestion (Der Marderosian *et al.*, 1964)." R. A. Heacock, "Psychotomimetics of the Convolvulaceae," *Progress in Medicinal Chemistry*, 11 (1975), pp. 112-113.

and several species of the genus *Argyreia.*[1] Its isomer, isolysergic acid hydroxyethylamide has also been identified in seeds of *R. corymbosa*, *I. violacea*, and a few species of *Argyreia.*[2]

[1] Jew-Ming Chao and Ara H. Der Marderosian, "Identification of Ergoline Alkaloids in the Genus *Argyreia* and Related Genera and their Chemotaxonomic Implications in the Convolvulaceae," *Phytochemistry*, 12 (1973), pp. 2435-2440; Schultes and Hofmann, *op. cit.*, pp. 151-152.

[2] Chao and Der Marderosian, "Identification of Ergoline Alkaloids," [Table 2].

Chapter 13: Lysergic Acid Hydroxyethylamide: Isomerization and Extraction

It is clear from the study conducted by Fajardo *et al.* that the levels of the toxic peptide alkaloids in *Claviceps purpurea*: ergocristine, ergocornine, α-ergokryptine, ergotamine, ergosine, and ergonovine (a water-soluble amide derivative of lysergic acid) are reduced during normal food processing, and the extent of this loss is related to the pH value of the food and temperature.[1] The method of boiling ergot was investigated by these authors using spaghetti processed with heavily contaminated ergot flour, dried at 39°C and 80°C and boiled for ~14 minutes. Following an initial alkaloid loss of 20-42%, thought to be due to an incomplete extraction caused by the coarseness of the ground uncooked spaghetti, the authors report: "Loss of alkaloids in the cooked spaghetti ranged from 42-79%, with higher losses being evident in the spaghetti dried at high temperature. The level of alkaloids detected in the cooking water was low regardless of the drying cycle used."[2]

The amides derivatives of lysergic acid and the toxic peptide alkaloids transform to their isomers under conditions related to pH, and this occurs more rapidly with higher temperatures. The isomerization that is of interest here occurs in acidic solutions, which converts the alkaloids in ergot to their pharmacologically inactive isomers. Acid-catalyzed isomerization also

[1] J. E. Fajardo *et al.*, "Retention of Ergot Alkaloids in Wheat During Processing," *Cereal Chemistry*, 72 (1995), pp. 291-298.
[2] Ibid., pp. 292, 296.

occurs in the corresponding dihydro derivatives of the peptide alkaloids, and these isomers are also pharmacologically inactive.[1] The boiling of ergot in water containing cereal would create the acidic conditions necessary for isomerization, as barley possesses a pH value of 5.19-5.32, and barley-malt is slightly more acidic.

The isomers of the alkaloids in ergot are very weak and almost inert, and are named ergocristinine, ergocorninine, α-ergokryptinine, ergotaminine, ergosinine, and ergometrinine.[2] To provide examples of isomerization, boiling ergocristine in a methyl alcoholic solution converts the alkaloid to its isomer ergocristinine.[3] Ergotamine slowly transforms to its isomer ergotaminine in alcoholic solution at room temperature, which occurs more rapidly when boiled in methyl alcohol.[4] Ergosine converts to ergosinine on acidification with sulfuric acid.[5] Isomerization with an acid catalyst causes some conversion of the lysergic acid alkaloid to isolysergic acid,[6] but also destroys lysergic acid.[7]

The amide derivatives of lysergic acid are also converted to their isomers with acid-catalyzed isomerization: ergometrine (ergonovine) converts to ergometrinine when heated in glacial acetic acid,[8] and ergine transforms to isoergine (erginine) when boiled in a mixture of alcohol and phosphoric acid.[9] Lysergic acid hydroxyethylamide readily transforms to isolysergic acid hydroxyethylamide in diluted acetic acid.[10] Indicating a certain degree

[1] H. Ott, A. Hofmann and A. J. Frey, "Acid-Catalyzed Isomerization in the Peptide Part of Ergot Alkaloids," *Journal of the American Chemical Society*, 88 (1966), pp. 1251-1256; Zdeněk Řeháček and Přemysl Sajdl, *Ergot Alkaloids: Chemistry, Biological Effects, Biotechnology*, Amsterdam, Elsevier, 1990, p. 44.

[2] Barger, "Alkaloids of Ergot," pp. 85-87.

[3] Ibid., p. 86.

[4] George Barger, *Ergot and Ergotism: A Monograph Based on the Dohme Lectures Delivered in John Hopkins University, Baltimore*, London, Gurney and Jackson, 1931, p. 134.

[5] Sydney Smith and Geoffrey Millward Timmis, "The Alkaloids of Ergot. Part VIII. New Alkaloids of Ergot: Ergosine and Ergosinine," *Journal of the Chemical Society* [London], 1937, pp. 396-401.

[6] Sydney Smith and Geoffrey Millward Timmis, "The Alkaloids of Ergot. Part VII. isoErgine and isoLysergic Acids," *Journal of the Chemical Society* [London], 1936, pp. 1440-1444.

[7] Barger, "Alkaloids of Ergot," p. 88.

[8] Sydney Smith and Geoffrey Millward Timmis, "The Alkaloids of Ergot. Part VI. Ergometrinine," *Journal of the Chemical Society* [London], 1936, pp. 1166-1169.

[9] Smith and Timmis, "isoErgine and isoLysergic Acids," p. 1442.

[10] F. Arcamone *et al.*, "Production of Lysergic Acid Derivatives by a Strain of *Claviceps paspali* Stevens and Hall in Submerged Culture," *Nature* [London], 187

of toxicity, lysergic acid hydroxyethylamide shows an estimated 30 to 50 percent of the pharmacological activity of ergometrine;[1] isolysergic acid hydroxyethylamide, isoergine, and ergometrinine are all pharmacologically inactive.[2]

The amide derivatives of lysergic acid have a lower melting point than the toxic peptide alkaloids, while the isomers of these alkaloids generally possess an equal or higher melting point, and it can be stated with certainty that none of these alkaloids would decompose from the heat of the boiling water, which of course boils at 100°C. The melting point of each alkaloid and its isomer is provided below, according to the *CRC Handbook of Data on Organic Compounds*:

Isoergine — 132°C.[3]
Ergine — 135°C.[4]
Lysergic acid hydroxyethylamide — 135°C.[5]
Ergonovine (Ergometrine) — 159°C.
Ergocristine — 175°C.
Ergocornine — 182-4°C.
α-Ergokryptine — 212°C.
Ergotamine — 213°C.
Ergosine — 228°C.[6]
Isolysergic acid hydroxyethylamide — 242°C.[7]

The isomers of the lysergic acid derivatives and toxic peptide alkaloids are estimated to contain about 1–3% of the activity of their corresponding alkaloids,[8] and all of these are to some extent alcohol-soluble: ergocorninine (m.p. 228°C) and α-ergokryptinine (245°C) are soluble in alcohol;

(1960), pp. 238-239.

[1] A. Glässer, "Some Pharmacological Actions of D-Lysergic Acid Methyl Carbinolamide," *Nature* [London], 189 (1961), pp. 313-314.

[2] Řeháček and Sajdl, *op. cit.*, pp. 39-40.

[3] *Dictionary of Natural Products*, London, Chapman & Hall, 1994, vol. III, p. 3719.

[4] Depending on the configuration of the amide substituent, ergine possess a melting point of 135°C and 242°C, and isoergine 132°C and 242°C. *CRC Handbook of Microbiology*, Second Edition, A. I. Laskin and H. A. Lechevalier (eds.), Boca Raton, Fl., CRC Press, Inc., 1977, vol. V, p. 616; *Dictionary of Natural Products*, vol. III, p. 3719; Smith and Timmis, "isoErgine and isoLysergic Acids," p. 1442.

[5] *Dictionary of Natural Products*, vol. III, p. 3719.

[6] *CRC Handbook of Data on Organic Compounds*, R. C. Weast and M. J. Astle (eds.), Boca Raton, Fl., CRC Press, Inc., 1985, vol. I, pp. 596-597.

[7] *Dictionary of Natural Products*, vol. III, p. 3719.

[8] Barger, "Alkaloids of Ergot," pp. 174, 192-193.

ergocristinine (237°C), ergotaminine (241°C),[1] and ergosinine (220°C) are very slightly soluble in alcohol;[2] and ergometrinine (196°C) is also soluble in alcohol.[3]

As Hofmann has stated, lysergic acid hydroxyethylamide, ergine, and ergometrine (ergonovine) are water-soluble alkaloids. But ergine is only slightly soluble in alcohol,[4] and this solubility would not increase in water, while its isomer isoergine is even less soluble in water,[5] but is soluble in alcohol.[6] Ergometrine is soluble in water, while its isomer ergometrinine is very sparingly soluble in water, but is soluble in alcohol.[7] Lysergic acid hydroxyethylamide shows some solubility in water, and its isomer isolysergic acid hydroxyethylamide is soluble in alcohol.[8] It would thus appear that the isomerization of the alkaloids, produced by the method of boiling in an acidic solution for a number of hours, would successfully detoxify *Claviceps* spp., and when the ergot is subsequently exposed to alcohol, the isomer isolysergic acid hydroxyethylamide can be extracted, producing an effective hallucinogenic potion – although consuming excessive quantities would remain dangerous.

Lysergic acid hydroxyethylamide occurs in both *Claviceps purpurea* and *Claviceps paspali*. However evidence has come forth that *Paspalum distichum*, which is infected by *C. paspali*, was not present in ancient Greece.[9] Hofmann writes: "I mention this only as a possibility or a likelihood, and not because

[1] "The solubility of ergotinine in alcohol at room temperature is 1:400, of ergotaminine 1:6400, by weight (Frèrejacque and Hamet)." Barger, *Ergot and Ergotism*, p. 138.

[2] *CRC Atlas of Spectral Data and Physical Constants for Organic Compounds*, vol. III, pp. 228-230.

[3] *CRC Handbook of Data on Organic Compounds*, vol. I, pp. 596-597; Smith and Timmis, "Ergometrinine," p. 1167.

[4] *CRC Atlas of Spectral Data and Physical Constants for Organic Compounds*, vol. III, p. 228.

[5] Smith and Timmis, "isoErgine and isoLysergic Acids," p. 1442.

[6] Jew-Ming Chao and Ara H. Der Marderosian, "Ergoline Alkaloidal Constituents of Hawaiian Baby Wood Rose, *Argyreia nervosa* (Burm. f.) Bojer," *Journal of Pharmaceutical Sciences*, 62 (1973), pp. 588-591.

[7] Smith and Timmis, "Ergometrinine," p. 1167.

[8] Chao and Der Marderosian, "Ergoline Alkaloidal Constituents," p. 590.

[9] Sheldon Aaronson, "Fungal parasites of grasses and cereals: their rôle as food or medicine, now and in the past," *Antiquity*, 63 (1989), p. 252; Francesco Festi and Giorgio Samorini, "*Claviceps paspali* and the Eleusinian *Kykeon*: A Correction," *Entheogen Review*, 8 (1999), pp. 96-97.

we need *P. distichum* to answer Wasson's question."[1] With the elimination of *P. distichum*, it appears that the hallucinogenic agent of the Eleusinian *kykeon* was obtained from *C. purpurea*, parasitic to barley and *Lolium temulentum*, the latter of which grows with the barley and was perceived by the Greeks as its avatar. Unfortunately, there is no way to determine whether *C. purpurea* has since undergone any adaptation and mutation, Hofmann writes: "We have no way to tell what the chemistry was of the ergot of barley or wheat raised on the Rarian plain in the 2nd millennium B.C. But it is certainly not pulling a long bow to assume that the barley grown there was host to an ergot containing, perhaps among others, the soluble hallucinogenic alkaloids."[2]

Hofmann analyzed the *C. purpurea* that grows abundantly on the wheat and barley in Greece, and found the expected peptide alkaloids, along with ergine, but not lysergic acid hydroxyethylamide.[3] Ergine only occurs in *C. paspali*, and does not occur in *C. purpurea*. Lysergic acid hydroxyethylamide is somewhat unstable and is hydrolyzed to ergine in the extraction procedure.[4] In a pH neutral or weak acid solution, lysergic acid hydroxyethylamide easily decomposes to ergine and acetaldehyde.[5] Although lysergic acid hydroxyethylamide (extracted as 0.4% ergine) is proportionally a small percentage of the total alkaloids contained in *C. purpurea*,[6] this comparison has little relevance after isomerization; very minute amounts of the drug will produce psychotomimetic effects if the epimerization of lysergic acid hydroxyethylamide produces an isomer possessing a fraction of the potency of LSD. However these facts remain: lysergic acid hydroxyethylamide has been identified in *C. purpurea* infecting cultivated cereals, and there is absolutely no doubt that ergot poisoning from these cereals can cause hallucinations.

[1] Hofmann, *Road to Eleusis*, p. 33.
[2] Ibid., p. 32.
[3] Ibid., p. 32.
[4] Albert Hofmann, "*Teonanácatl* and *Ololiuqui*, two ancient magic drugs of Mexico," *Bulletin on Narcotics*, 23 (1971), p. 9; Heacock, *op. cit.*, p. 97.
[5] Albert Hofmann, *Die Mutterkornalkaloide*, Stuttgart, Ferdinand Enke, 1964, p. 34; A. Stoll and A. Hofmann, "The Ergot Alkaloids," *The Alkaloids: Chemistry and Physiology*, R. H. F. Manske (ed.), New York, Academic Press, 1965, vol. VIII, p. 747.
[6] Aaronson, *Fungal parasites*, p. 252.

Chapter 14: The *Kykeon* and the Dionysian Potion of the Lesser Mystery

Calvert Watkins has convincingly demonstrated that the Soma beverage was formulaically identical to the *kykeon* of the Eleusinian rites, and concludes that Homer's references to barley and honey are to ergot-infected barley and honeydew.[1] In Homer's *Iliad*, Athena is offered libations of "honey-sweet wine,"[2] and in the *Odyssey*, Circe's potion contains barley and drugs mixed in wine: "[She] made for them a potion of cheese and barley meal and yellow honey with Pramnian wine; but in the food she mixed baneful drugs."[3] But in this formula provided by Homer, the addition of drugs to the wine occurs *after* the wine has already been prepared, which suggests that the ergot-infected barley has already been rendered non-toxic, an absolute necessity considering that the toxic peptide alkaloids are alcohol-soluble. Perhaps the most revealing quote is found in the *Iliad*:

> Hecamede mixed a potion...pale honey, and ground meal of sacred barley; and beside them a beauteous cup...Therein the woman, like to the goddesses, mixed a potion for them with Pramnian wine, and on this she grated cheese of goat's milk with a brazen grater, and sprinkled thereover white barley meal; and she

1 Watkins, "Praise Famous Grains," pp. 9-17.
2 Homer, *Iliad* X.579.
3 Homer, *Odyssey* X.233-236.

bade them drink, when she had made ready the potion.[1]

In this identical formula, Homer makes reference to ergot: "pale honey, and ground meal of sacred barley," which is mixed in the Pramnian wine, and like the Soma beverage this is sprinkled with milk product and barley meal. Under these conditions described by Homer, the ergot alkaloids contained in the "ground meal of sacred barley" have previously underwent isomerization; otherwise this potion would clearly be poisonous due to the alcohol-soluble peptide alkaloids.

According to the *Homeric Hymn to Demeter*, the only ingredients of the *kykeon* were barley, water, and *glechon*, which is provided to the Goddess Demeter while in search of her daughter Persephone: "Metaneira offered her a cup filled with wine, as sweet as honey, but she refused it, telling her the red wine would be a sacrilege. She asked instead for barley and water to drink mixed with...*glechon*."[2] Although roasting isn't necessary to produce malt, the Greeks soaked barley in water and dried this overnight, then roasted it and crushed this between stones.[3] In Ovid's *Metamorphoses*, while searching for her daughter in Sicily, the Goddess consumes a potion containing roasted barley-groats, which, as Karl Kerényi concludes,[4] is clearly an ingredient of the *kykeon*: "Then out came an old woman and beheld the goddess, and when she asked for water gave her a sweet drink with parched barley."[5] In the formula provided in the *Homeric Hymn to Demeter*, barley clearly refers to the malted barley that produced the fermentation of the liquid, although this does not preclude the further addition of parched barley to the mixed potion.

The Greek term *glechon* (or *blechon*) is translated as "pennyroyal," and like all other botanical references in the Greek and Latin texts, it is imperative that this term is considered and defined in the context of pre-Linnaean botany and the Hippocratic tradition to which it belongs. According to Pliny, *glechon* was cultivated and was a highly effective drug used in Hippocratic medicine,[6] a homeopathic medical tradition that cures by similitude, sometimes referred to as the "hair of the dog," in which the cause of the disease is also the source

[1] Homer, *Iliad* XI.624-641.
[2] *Homeric Hymn to Demeter* 206-210.
[3] Pliny, *Natural History* XVIII.72.
[4] Kerényi, *Eleusis*, p. 178.
[5] Ovid, *Metamorphoses* V.448-450.
[6] Pliny, *Natural History* XX.152-157.

of the cure. Preceding the allopathic medical tradition that cures by contraries and follows Aristotle and Galen, Hippocratic medicine was intimately connected with the religious institutions of the ancient Greeks, evidenced by the gods invoked in the Hippocratic Oath.[7] This particular topic is beyond the scope of the present work, and I have returned to this subject in a book devoted to the Scientific Revolution and the Great Plague in London in 1665.[8]

Grapes possess an acidic pH value of 3.5–4.5, which provides the necessary acidity to produce the isomerization of the alkaloids in ergot. In his work *On Agriculture*, Columella provides the formulaic instructions for producing *glechonites*, or "pennyroyal wine," in which one Roman pound (approximately 11.5 ounces) of *glechon* is boiled in four *sextarii* (*sextarius* = approximately 1.15 pints) of grape juice (must) until a quarter of the original quantity remains, and is then allowed to ferment:

> The following is the way to make up wines flavoured with wormwood, hyssop, southern-wood, thyme, fennel and pennyroyal. Take a pound of Pontic wormwood and four *sextarii* of must and boil down to a quarter of the original quantity and put what remains when it is cold into an *urna* of Aminean must. Do the same with the other things mentioned above...it is called the pennyroyal brand of wine [*glechonites*].[9]

The secret ingredient identified as *glechon* in the *Homeric Hymn to Demeter*, which Homer calls the "ground meal of sacred barley," was mixed with two other ingredients in the *kykeon*: water, and barley that had been soaked, dried, roasted, and ground between stones. These two ingredients — water and roasted barley malt — produced the fermentation of the *kykeon*. The evidence also suggests that in the preparation of the *kykeon*, the ingredient called *glechon*, the ergot-infected "ground meal of sacred barley," was first boiled, dried, and ground between stones, and was later mixed into the fermented liquid during the ceremony. This method conforms to the formulas provided in Homer's *Iliad* and *Odyssey*, and the *Homeric Hymn to Demeter*: "barley and water...mixed with...*glechon*." Carl A. P. Ruck writes:

> Barley-groats, water...and *glechon* (also called *blechon*). These

[7] Shelley, *Elixir*, pp. 252-260.

[8] Scott Shelley, *Science, Alchemy and the Great Plague of London*. New York, Algora Publishing, 2017.

[9] Columella, *On Agriculture and Trees* XII.35.

> are the known ingredients for the Eleusinian potion, the mixed drink or *kykeon*...The only supposed evidence for the mixing of the *kykeon* in advance is a papyrus fragment of a comedy of Eupolis where a foreigner is seen with barley-groats (*krimnon*) on his upper lip while still in Athens; the circumstances, however, are probably the affair of the Profanation of the Mysteries, which involved exactly this, the drinking of the *kykeon* illegally at home as a recreational inebriant, since certain well-placed Athenians with connections to the Eleusinian priesthood had apparently learned the secret of the potion. And indeed, Eupolis's foreigner is described as being overtaken by a hallucinatory fever (*epialos*, like a nightmare) while on the way to the marketplace because of the *kykeon* he had drunk...Inside [the temple of Eleusis], but certainly not the exaggerated thirty thousand of them as has been claimed, for the Hall could accommodate perhaps only a thirtieth of that, they ranged themselves along the peripheral steps, sitting probably, not standing all packed together, and watched as the priestesses danced in the darkness with lanterns and incense burners on their heads. The Mystery Baskets or Cista Mystica were opened, whatever they contained hidden until now from profane view. Then the *kykeon* was mixed...One of the Eleusinian priests called the Hydranos was specifically responsible for the water, for this would have to be prepared in advance..."I have opened the basket; I have drunk the potion," was the final password.[1]

The revolutionary advances in ancient technology, particularly in agriculture, may contain the answer to the origination of the Eleusinian *kykeon* and its Vedic counterpart Soma. Barley and wheat are the universal cereals of Old World plant domestication, and were the founder crops of Neolithic agriculture.[2] Barley was a universal companion of wheat, but in comparison was considered to be an inferior staple, and was the main cereal used for fermentation in the Old World.[3]

The basis of all beer production is the fermentation of starch

[1] Ruck, "Mixing the Kykeon," pp. 77-80.
[2] Renfrew, *op. cit.*, pp. 149-172; Zohary and Hopf, *op. cit.*, pp. 19-69.
[3] Zohary and Hopf, *op. cit.*, p. 59.

> in amylaceous cereals. Grain always contains a small quantity of directly fermentable sugar, but this is inadequate in amount to produce an alcoholic drink. As starch, itself, cannot ferment unless first split into fermentable sugars, it is usually first subjected to malting; i.e. letting grain germinate, a process during which starch is converted into maltose, and considerable amounts of diastase are developed. In modern processes, malt is then heated and dried to stop germination, then boiled with water, strained and incubated with yeast.[1]

R. J. Forbes conjectured that the knowledge of fermentation may even be older than agriculture, although it does appear certain that the knowledge of malting barley led to this discovery, since there is little, if any evidence for the use of bee's honey during this time:

> Fermentation needs fire and pottery. Though simple forms of pottery were known by the end of the Upper Palaeolithic the plants collected in those days gave sweet infusions seldom suitable for fermentation, as in general they contain too little sugar. The techniques of fermenting came with organized agriculture, some traces of which go back to the Upper Palaeolithic...However a more regular production of cereals suitable for fermentation came in Neolithic times only. Even the wild species of grapes, berries and other fruit collected and eaten in those days would hardly make a suitable base material for fermentation...Mead, the fermented drink made from honey (and grain) may even be older than agriculture, but this we will probably never know...Nectar and ambrosia were probably mead-like concoctions.[2]

The early discoveries that led to the practice of domesticating cereal plants, which resulted in the agricultural revolution of the Neolithic period, was a technological milestone that preceded the advent of horticulture by several thousand years. The earliest archaeological evidence for viticulture comes from the eastern Mediterranean, and this material has been dated to the Early Bronze Age.[3] The technology involved in fermenting beverages using either

[1] Darby, Ghalioungui, and Grivetti, *op. cit.*, vol. II, p. 534.
[2] Forbes, *op. cit.*, vol. III, p. 60.
[3] Zohary and Hopf, *op. cit.*, pp. 151-159.

cereal or fruit as the base material is the same; it therefore seems reasonable to assume that the knowledge of fermenting cereals was simply applied to the cultivated grape in making wine, and the water was replaced with grape juice to which the cereal ingredients continued to be added. The Greek term for wine, *μέθυ* (*méthu*), is etymologically identical to Old English *medu*, "mead," and Sanskrit *mádhu*,[1] a synonym for Soma that is variously defined in the Indo-European languages as honey, mead, or wine.[2] In addition, there exists a great deal of evidence that the alcoholic preparations of the Greeks and Romans were extremely dangerous when consumed in excess. This is revealed by the various symptoms and diseases reportedly caused by wine in the Greek and Latin texts. For instance, in his translation from the Greek to Latin of the medical works of Soranus of Ephesus, Caelius Aurelianus attributes to excessive wine drinking a condition called "erysipelas," a name commonly used for ergot poisoning.[3]

On the banks of the Ilissos near Athens stood the sanctuary of Agrai, where the Lesser Mystery was performed in early spring in preparation for the Greater Mystery performed at Eleusis in the month of September. This preliminary rite was intimately connected with the Dionysian festivals held in February, and Kerényi provides convincing evidence that this ceremony involved Dionysian wine, Dionysus of course being the "god of wine," but also the "god of *all* inebriants."[4] The Hellenic Greeks and the Romans cultivated the poppy and produced opium,[5] and Kerényi's work leaves little doubt that the Lesser Mystery of the Eleusinian rites involved not only wine,

[1] Liddell and Scott, *op. cit.*, p. 1091.

[2] Macdonell and Keith, *op. cit.*, vol. II, pp. 123-124; Watkins (ed.), *Dictionary of Indo-European Roots*, p. 52.

[3] Caelius Aurelianus, *On Acute Diseases and On Chronic Diseases*, I. E. Drabkin (trans.), Chicago, University of Chicago Press, 1950, pp. 301-303.

[4] C. Kerényi, *Dionysos: Archetypal Image of Indestructible Life*, R. Manheim (trans.), Princeton, Princeton University Press, 1976, pp. 52-125; Kerényi, *Eleusis*, pp. 51-52, 64.

[5] Kritikos and Papadaki, *op. cit.*, pp. 5-10, 17-38; Mark David Merlin, *On the Trail of the Ancient Opium Poppy*, Rutherford, Fairleigh Dickinson University Press, 1984, pp. 218-250.

but also grain[1] and opium.[2]

The Cretans knew the method of incising the poppy head to produce opium as early as Late Minoan III, and used opium as a psychoactive agent in their religious rites.[3] The islands of Crete and Cyprus were settled outposts of the Minoan-Mycenaean cultures that flourished throughout the Aegean in pre-Hellenic times.[4] The inhabitants of Cyprus exported their opium in containers shaped in the form of poppy capsules, which have since been designated as Base-ring I juglets.[5] These containers have tested positive for opium, which apparently was exported in liquid form, dissolved in either water or wine.[6] This method of dissolving opium in liquid creates the problem of mold, and is counter-productive from an economic standpoint. However, it certainly seems within the realm of possibilities that they first

[1] "As would be expected, two crops were planted. The times of planting do not correspond exactly to the two Mysteries, which appear to be religious preliminaries for the actual plantings. Hesiod's *Work and Days* directs that the winter crop of Demeter's grain (presumably barley) should be planted when the voice of the migratory crane is first heard, *i.e.*, mid November (448)...The (summer) crop should be planted when the cuckoo first calls, *i.e.*, March (486)...This crop is winnowed when Orion first rises at dawn, *i.e.*, July (587)." Ruck, "Mixing the Kykeon," p. 85.

[2] Kerényi, *Eleusis*, pp. 55-57, 120-144, 158-168.

[3] Kritikos and Papadaki, *op. cit.*, p. 25.

[4] David G. Hogarth, *Ionia and the East; Six Lectures Delivered before the University of London*, Oxford, Clarendon Press, 1909.

[5] Robert S. Merrillees, "Opium Trade in the Bronze Age Levant," *Antiquity*, 36 (1962), pp. 287-292; Merrillees, "Opium again in Antiquity," *Levant*, 11 (1979), pp. 167-171; Merrillees, "Highs and Lows in the Holy Land: Opium in Biblical Times," *Eretz-Israel*, 20 (1988-1989), pp. 148-155; Andrew Sherratt, "Sacred and Profane Substances: the Ritual Use of Narcotics in Later Neolithic Europe," *Sacred and Profane: Proceedings of a Conference on Archaeology, Ritual and Religion. Oxford, 1989*, P. Garwood, D. Jennings, R. Skeates and J. Toms (eds.), Oxford, Published by the Oxford Committee for Archaeology, 1991, p. 56.

[6] "In a classic paper in ANTIQUITY thirty years ago, Mr de Navarro discussed the trade relations between Massilia and the Celtic world of Early Le Tène times. He pointed out that among the objects of Mediterranean and classical origin finding their way into Central Europe from the 6th century B.C. onwards, 'most of the imported vessels, whether pottery or bronze, were ultimately connected with the carrying, serving, mixing and drinking of wine'—and in fact he went further, and in a now famous phrase roundly declared that 'La Tène art may largely have owed its existence to Celtic thirst'...Wine, was, after all, an ancient drink in Egypt and the Levant at least, and seems to have been a Mycenaean product, and it is tempting to think that the new bronze vessels manufactured and traded in Continental Europe from the 12th century B.C. may not only have had a technical link with the Aegean or Levantine world in their mode of manufacture, but have owed their popularity at least in part, to the civilizing qualities of their contents. Perhaps not only La Tène art, but the achievement of Late Bronze Age Europe, may owe something to thirst." Stuart Piggott, "A Late Bronze Age Wine Trade?" *Antiquity*, 33 (1959), pp. 122-123.

boiled ergot-infected grain in grape juice, and later added opium after fermentation, creating a potent mixture containing alcohol, opium, and isolysergic acid hydroxyethylamide — but until a method of testing for the presence of ergot in archaeological artifacts is developed, this hypothesis will remain unchallenged. The extremely potent and potentially lethal Dionysian wine, which according to Homer was often diluted with water in a ratio of 20:1,[1] was intimately connected with Greek religion and commonly used in religious ceremonies, and it seems reasonable to conclude that it was also employed in the Lesser Mystery of the Eleusinian rites, and this is supported by a body of evidence.[2]

[1] Homer, *Odyssey* IX.208-211; in Murray, *Odyssey*, vol. I, p. 317.

[2] Neuburger, *op. cit.*, pp. 100-109; Kerényi, *Eleusis*, pp. 51-52, 64, 138; Carl A. P. Ruck, "The Wild and the Cultivated: Wine in Euripides' *Bacchae*," *Journal of Ethnopharmacology*, 5 (1982), pp. 231-270; Wasson *et al.*, *Road to Eleusis*, pp. 85-98; William Scott Shelley, *The Origins of the Europeans: Classical Observations in Culture and Personality*, San Francisco, International Scholars Publications, 1998, pp. 95-182.

Chapter 15: Ergot Poisoning

The mixing of the *kykeon* is a topic belonging to the history of science that includes the subjects of classical literature, chemistry, pharmacology, and also epidemiology. Extending his hypothesis to its logical conclusion, Webster has suggested that partial hydrolysis was responsible for the outbreaks of ergot poisoning that produced profound psychological symptoms, and that this process was absent from the epidemics in which these symptoms were not present. Webster writes in an email dated 2 March 2003 received by Carl Ruck:

> Not to be overlooked is the possibility that fermentation of sorts might also bring about partial hydrolysis, since as Kren found, simple bacteria can do the job. Perhaps the enzymes of yeasts can also do so. Even if a "fermentation" occurred in basic solution, then of course some partial hydrolysis could occur.

This is inaccurate. Partial hydrolysis occurs in alkaline solutions, and alcohol is acidic, which causes the isomerization of the alkaloids in ergot.

> Baking alone can only decompose the alkaloids to inactive breakdown products. However, as above, it would be the presence of basic material, or bacteria, in the flour, the bread dough, etc., that BEFORE BAKING, might have produced some degree of partial hydrolysis to ergine/isoergine.

This is reference to Fajardo *et al.*, who report the following regarding bread baked at 225°C for 25 minutes: "bread baked from the HEC (heavily ergot contaminated) clear flour exhibited no loss of alkaloids in the crumb. However, alkaloid losses of 25–55% were evident in the crust. Lower concentrations of ergot alkaloids in the crust are likely due to higher crust temperatures during baking."[1] These alkaloid losses in the crust are clearly the product of acid-catalyzed isomerization and not partial hydrolysis. However, under the right conditions, some degree of partial hydrolysis might very well occur, and if ergine and isoergine were hallucinogenic, one would expect to find evidence for this in epidemics of ergot poisoning. Webster writes:

> Obviously, for ergotism, only a portion of the ergotoxine alkaloids would have been converted, whereas for the kykeon, a nearly complete conversion. As for tripping whilst one's limbs are falling off, it would not be uplifting, and medieval folk who were 'hallucinating' on ergine/isoergine whilst in the throes of vasoconstrictive agony can [be] excused for not realising they were seeing the white light...

Hallucinations do not occur with the vasoconstrictive type of ergot poisoning, the form known as gangrenous ergotism. Hallucinations only occur with convulsive ergotism. Mr. Webster is obviously confusing the two distinct types of ergot poisoning, which are examined below. He writes:

> As for ergotism, I am going to introduce a new hypothesis at the conference: that various ways of making bread, and possible infection of the flour and bread-dough with certain bacteria, might have led to some partial hydrolysis and thus the outbreaks of ergotism differed greatly in characteristics, thus explaining why some outbreaks were merely vasoconstrictive disease and gangrenous, and others involved distinct psychoactive, hallucinatory experience. If bread were made from flour that had been ground on mills leaving a basic residue from the stones, or sour-dough bread allowed to ferment before baking...Remember Kren said that even common soil bacteria can bring about the partial hydrolysis.

Ergot poisoning manifests in two forms: gangrenous and convulsive. Epidemics of either gangrenous or convulsive ergotism typically recurred within a specific geographical region. For instance, in England the disease

[1] Fajardo *et al.*, *op. cit.*, pp. 292, 296.

took the convulsive form throughout history, and gangrenous ergotism occurred here only once, which attacked the family of a poor agricultural laborer in Wattisham in 1762.[1] Referring to his classic monograph *Ergot and Ergotism*, George Barger writes:

> Most mixed epidemics appear to have occurred in Russia; generally however nervous ergotism predominated in the north, while the gangrenous type has been recorded from the south of that country. In France gangrenous ergotism was the rule and of 41 references to ergotism, which I have collected from French chronicles, only two mention convulsions. Both references refer to the same epidemic (or to two epidemics in rapid succession) in Lorraine, bordering on Germany, where convulsive ergotism was the universal type. A small Swiss epidemic of 1709, accurately described by LANG, is often quoted as an example of the mixed type, but was essentially gangrenous...In Germany gangrene was extremely rare. The chronicle of Meissen in Saxony for 1486 mentions it. Apart from a few sporadic cases, such as that of BRUNNER, there was no ergot gangrene in Germany (TAUBE is very definite on this point); none is recorded for Bohemia, Hungary, Sweden or Finland. East of the Rhine gangrene was only met with in a few of the many Russian epidemics, in which convulsive symptoms greatly predominated...[2]

Gangrenous ergotism and convulsive ergotism share a number of milder symptoms at the early stages of the disease, but rarely do both forms occur together.[3] In gangrenous ergotism the symptoms generally remain mild until a critical level is reached and gangrene sets in, and individual susceptibility to gangrenous ergotism varies greatly, and rarely are entire families attacked by it (in Wattisham the father escaped). Mania and hallucinations are absent in the gangrenous form, and are only present in convulsive ergotism, a violent disease that commonly affects whole families, and consequently was considered to be contagious.[4] Barger writes:

[1] Charles Creighton, *A History of Epidemics in Britain*, Cambridge, Cambridge University Press, 1891, vol. I, pp. 59-63; Barger, *Ergot and Ergotism*, pp. 27-28, 63-64.
[2] Barger, "Alkaloids of Ergot," pp. 205-206.
[3] Barger, *Ergot and Ergotism*, pp. 20-23, 80-81.
[4] Ibid., pp. 21, 27, 37, 59.

> The two Types of Ergotism, the gangrenous and the convulsive, are only sharply differentiated in severe cases. A number of early and mild symptoms are common to both...The early symptoms in sub-acute ergotism include a general lassitude, depression, vague lumbar pains, and pains in the limbs, particularly in the calf, nausea, occasionally vomiting; all these have been observed in human subjects after an intravenous or subcutaneous injection of 0.5 mg. ergotamine or ergotoxine. Often the intellect was dulled, and this also has been recorded after larger "clinical" doses of the alkaloids in man...A characteristic tingling of the skin (formication)...[is] closely associated with the convulsive type...Yet in many cases formication preceded gangrene in the French epidemics of 1814, 1816 and 1820...This symptom occurred in the case of ergotamine poisoning described by Carreras.[1]

In severe cases of gangrenous ergotism, those afflicted with the disease often report feeling "almost drunk," although in "some years the ergot was not found to have any harmful effects."[2] Barger writes: "In cases of recovery from the gangrene the patients remained dull and stupid for the rest of their lives."[3] In *The Alkaloids of Ergot*, Barger describes the disfigurement of the victims of gangrenous ergotism:

> Apart from the above early and mild symptoms, probably common to both types, severer and more distinctive effects appeared after some days or weeks, with swelling and inflammation of the extremities; often only one part was attacked, more frequently a foot than a hand. This was followed by extreme pain; the shrieks of the sufferers alarmed their neighbours in an English outbreak in 1762. There was at first a feeling of heat...Later a feeling of intense heat alternated with one of cold...Gradually the part affected became numb, the pains sometimes stopped suddenly...Later the diseased part became black...often quite suddenly, and all sensation was lost. The gangrenous part shrank, became mummified and dry; the whole

[1] Barger, "Alkaloids of Ergot," p. 198. Ergotoxine is a mixture of ergocornine, ergocristine, and ergokryptine. See also, E. A. Cameron and E. B. French, "St. Anthony's Fire Rekindled: Gangrene due to Therapeutic Dose of Ergotamine," *British Medical Journal*, 2 (1960), p. 28.

[2] Barger, *Ergot and Ergotism*, pp. 59-60.

[3] Ibid., p. 61.

> body was emaciated and the gangrene gradually spread upwards; sometimes there was putrefaction (moist gangrene). In severe cases the course of the disease was much more rapid; with violent pains for 24 hours as the only premonitory sign, gangrene might set in suddenly. The separation of the gangrenous part often took place spontaneously at a joint without pain or loss of blood...In bleeding their patients the surgeons found it difficult to obtain a satisfactory flow of blood. The extent of the gangrene varied from the mere shedding of nails and the loss of fingers or toes...to the loss of all four limbs. After the loss of a single limb...the patient often made as good a recovery as from a modern amputation, and lived for many years...The absence of severe permanent damage (other than the loss of a limb) rather sharply differentiates the gangrenous type from the convulsive which often caused irreparable lesions in the central nervous system.[1]

In contrast with the gangrenous form of ergot poisoning, in 1746 and the following year a widespread epidemic of convulsive ergotism occurred in Sweden, Russia, and the European mainland, in which the "pain was most violent; so that the fire in the limbs drove the victims hither and thither — some in their agony hurling themselves against walls or even into the water. Those grievously attacked generally died; those who survived became blind, dumb, or demented."[2] Barger identifies a "Malignant Fever" as convulsive ergotism in Daniel Sennert's *Of Agues and Fevers*, translated and published in 1658:

> It seized upon men with a twitching and kind of benummedness in the hands and feet, sometimes on one side, sometimes on the other, and sometimes on both: Hence a Convulsion invaded men on a sudden when they were about their daylie employments, and first the fingers and toes were troubled, which Convulsion afterwards came to the arms, knees, shoulders, hips, and indeed the whole body, until the sick would lie down, and roul up their bodies round like a Ball, or else stretch out themselves straight at length: Terrible pains

[1] Barger, "Alkaloids of Ergot," pp. 198-199.

[2] Clifford Allbutt and Humphry Davy Rolleston (eds.), *A System of Medicine*, London, Macmillan and Co., Limited, 1908, vol. II, pt. I, p. 887.

> accompanied this evil, and great clamours and scrietchings did the sick make; some vomited when it first took them. This disease sometimes continued some days or weeks in the limbs, before it seized the head, although fitting medicines were administered; which if they were neglected, the head was then presently troubled, and some had Epilepsies, after which fits some lay as it were dead six or eight hours, others were troubled with drowsiness, others with giddiness, which continued till the fourth day, and beyond with some, which either blindness or deafness ensued, or the Palsie: When the fit left them, men were exceeding hungry contrary to nature; afterwards for the most part a looseness followed, and in the most, the hands and feet swell'd or broke out with swellings full of waterish humours, but sweat never ensued. This disease was infectious, and the infection would continue in the body being taken once, six, seven, or twelve moneths. This disease had its original from pestilential thin humours first invading the brain and all the nerves; but those malignant humours proceeded from bad diet when there was scarcity of provisions. This disease was grievous, dangerous, and hard to be cured, for such as were stricken with an Epilepsie, were scarce totally cured at all, but at intervals would have some fits, and such as were troubled with deliriums, became stupid. Others every yeer in the month of *December* and *January*, would be troubled with it.[1]

Barger also writes of convulsive ergotism:

> In severer cases, the whole body was attacked by general convulsions, often so suddenly "that some at the table dropped knife or spoon and sank to the floor, and other fell down in the fields while ploughing"...If not confined to bed, the sufferers "tumbled about as if drunk"...The loud cries of the sufferers are often referred to in a graphic manner. A cold sweat covered the whole body and the spasm of the abdominal muscles caused violent retching. Occasionally the disease first showed itself by convulsions, two or three days after

[1] Daniel Sennert, *Of Agues and Fevers. Their Differences, Signes, and Cures. Divided into four Books: Made English by* N. D. B. M. *late of* Trinity *Colledge in* Cambridge, London, Printed by *J. M.* for *Lodowick Lloyd*, at the Castle in Cornhil, 1658, pp. 114-115; Barger, *Ergot and Ergotism*, pp. 32-33.

> eating the poisonous bread...In extreme cases the patients would lie for six or eight hours as if dead; in the 1597 epidemic some narrowly escaped being buried alive (Marburg). In such cases there followed a pronounced anæsthesia of the skin, the lower limbs became paralysed, and the arms subject to violent jerky movements; epileptiform convulsions, delirium, imbecility, and loss of speech were apt to occur in such patients, who became unconscious and generally died on the third day after the onset of the first symptoms. In severe but non-fatal cases, the disease might last for six to eight weeks, and convalescence took several months...Relapses were frequent...These relapses were accompanied by epilepsy, hemiplegia, and paraplegia (von Leyden). Among the after-effects of a severe attack may be mentioned: general weakness, trembling of the limbs, gastric pains, chronic giddiness, permanent contractures of the hands and feet, anæsthesia of finger and toes, impairment of hearing and of sight, and various mental derangements...The effects on the mind consisted of dullness and stupidity, even in the less severe cases...the more general disturbances in severe cases was dementia...Minor nervous defects, spasms and a dull intellect may persist for a long time in the adult, and serious relapses occurred years after the first attack...Psychoses due to ergot have been especially studied by Gurewitsch (1911) and von Bechterew (1892). A graphic early description of a patient with delusional insanity, seven months after the harvest, was already given by Hoffmeyer (1742); the constant movements of the hands and feet were only interrupted by tetanic convulsions.[1]

Mania and hallucinations are symptoms that only occur with convulsive ergotism, a form of the disease does not occur in France, just as gangrenous ergotism rarely occurred outside of France and Southern Russia. This raises the question: why didn't these particular psychological symptoms ever occur in France if partial hydrolysis causes the toxic alkaloids in ergot to become hallucinogenic? In the same line of reasoning, can the symptoms of convulsive ergotism be explained by partial hydrolysis? Considering the hard scientific facts, one can only conclude that the partial hydrolysis

[1] Barger, *Ergot and Ergotism*, pp. 34-37.

hypothesis fails to meet not only the crucial pharmacological evidence, but the epidemiological evidence as well.

Conclusion

It is evident that Soma and Surā are intimately connected in the Vedic literature, and the ingredients and methods that were used to prepare both beverages are nearly identical. Considering the evidence, it appears conclusive that the Soma liquor contained alcohol produced from malted barley (*i.e.* Surā), and the Surā liquor in the Caraka Sautrāmaṇī and the performance of this rite in the Rājasūya also contained Soma. It is also clear from the literature that Soma was cooked in cereal foods and consumed in solid form. The ingredients and methods used to prepare Soma, the production of large quantities of the Soma beverage and its frequent consumption, and the continual use of the genuine Soma-plant for centuries in India are physical parameters that necessarily determine the viability of any candidate for Soma, and only ergot-infected cereal meets all of these requirements.[1] The praise of Soma by the Vedic priests is also conferred upon barley, which is

[1] "In my opinion, *Amanita muscaria* is unsuitable for any identification with *soma/haoma* on the following grounds...*soma/haoma* is prepared from stems or stalks, which most probably should be regarded as fibrous (Brough 1971; Falk 1989) while the fleshy stems of *A. muscaria* contain only very small amounts of the pharmacologically active compounds, which are concentrated instead in the mushroom cap (these are the only parts of the mushroom used in northern Siberia)...culturally, the use of *A. muscaria* occurs only among the shamanistic peoples of northern Eurasia and it is neither a required part of any shamanistic rite, nor regarded as holy in them...the mushroom must have been rare in any of the proposed Indo-Iranian homelands. In contrast, when the use of *soma/haoma* began, the Aryans seem to have been inhabiting a region where the to-date unidentified plant was abundant." Nyberg, *op. cit.*, 392-393.

granted a mythological status in the Vedic literature, further evidence that the god-like "mighty barley" is the genuine Soma-plant of the *Ṛgveda*, and this identification is explicitly supported by the Vedic texts. It stands to reason that in the preparation of Soma, the ergot-infected barley, abundant with honeydew, was malted or mixed with malt, boiled, fermented, crushed, and then filtered, producing a kind of beer or mead containing the psychoactive alkaloid isolysergic acid hydroxyethylamide. Among the other enigmatic traditions of the ancient Indo-Europeans, the Athenian priests that performed the religious rites of the Eleusinian Mysteries would have certainly employed a similar method of producing a psychoactive potion from ergot.

It is clear from the physical evidence that the vessels excavated from the temples of Margiana contain the remains of cultivated cereals, specifically broomcorn millet, a host to *Claviceps purpurea.* The ergot-infected millet (with malt) was subjected to prolonged boiling, then allowed to ferment in large vessels, after which the cereal was removed from the fermented liquid, ground with stone implements, and was then strained with the fermented liquid through a wool filter placed on the bottom of the earthenware strainers. This is a method consistent with the preparation of Soma in the Vedic texts, and there are also similarities to the method employed by the Eleusinian priesthood to produce the *kykeon*, or "mixed drink."

Sarianidi's multi-plant Soma/Haoma hypothesis of Ephedra, opium, and hemp is based upon the interpretation of the impressions in the vessels as being those of cannabis seeds and Ephedra stems, and the tests performed by Meyer-Melikyan and Avetov. These tests do not appear to be reliable, as Harry Nyberg tested samples from Tokolok-21 and Gonur I and found neither Ephedra nor poppy pollen, but pollen which may belong to grain crops. Furthermore, the impressions interpreted to be cannabis seeds and Ephedra stems are clearly not accurate, as Bakels, Bottema, Cappers, Nesbitt, and Fuller have identified these impressions as broomcorn millet, and the tests performed by Bakels reveal no pollen of Ephedra, poppy, or hemp. There also currently exists no evidence that the Indo-Aryans of Central Asia were familiar with the cultivation of *Papaver somniferum*, or had any direct knowledge of *Cannabis sativa* in the second millennium B.C. There is no mention of either of these plants in the *Ṛgveda.*

The Indo-Aryans of the Bactria-Margiana Archaeological Complex were agricultural tribes that were undoubtedly familiar with the fermentation of cereals. There is absolutely no doubt that the fermentation of cereals was the function of the ceramic vessels excavated from the temples of Margiana, which was unlikely only for the alcohol, a theory few take seriously. The simplest and most logical explanation of the existing evidence of the identity of the Soma/Haoma libation produced in the temples of Margiana lead directly to ergot-infected cereal. The method was simple. The ergot-infected cereal was boiled for a prolonged period, causing the acid-catalyzed epimerization of the alkaloids in ergot due to the acidic pH of the cereal, and this was followed by the fermentation and consequently the alcohol-extraction of isolysergic acid hydroxyethylamide. The only amide derivatives of lysergic acid contained in *Claviceps purpurea* are lysergic acid hydroxyethylamide and ergonovine, and the isomer of the latter can be ruled out as the hallucinogenic agent produced from cooking ergot with the host cereal.

In summary of the chemistry, the evidence contradicting the partial hydrolysis hypothesis is substantial, while the scientific evidence supports (and perhaps more importantly, does not contradict) the acid-catalyzed isomerization and alcohol extraction method that has been proposed in the preceding pages. The prolonged boiling of *Claviceps purpurea* in an acidic solution created by the host-plant converts the amide derivatives of lysergic acid and the toxic peptide alkaloids to their pharmacologically inactive isomers, and the proposed hallucinogenic agent, isolysergic acid hydroxyethylamide, can then be extracted from the fungus with ethyl alcohol, produced by the technologically simple process of fermentation. The ergot-infected grain can also be boiled, dried, ground, then made into bread and other cereal foods, or mixed in wine or fermented barley — evidently the method of preparing the *kykeon* — and these products would also be hallucinogenic. Because of the acidic nature of cereals used in food processing, and the readiness of lysergic acid hydroxyethylamide to convert to its isomer (or decompose to ergine with a weak acid or alkaline catalyst), hallucinations in convulsive ergotism would appear to be caused by the epimerization of lysergic acid hydroxyethylamide.

Bibliography

Aaronson, Sheldon, "*Paspalum* spp. and *Claviceps paspali* in Ancient and Modern India," *Journal of Ethnopharmacology*, 24 (1988), pp. 345-348.

Aaronson, Sheldon, "Fungal parasites of grasses and cereals: their rôle as food or medicine, now and in the past," *Antiquity*, 63 (1989), pp. 247-257.

Aelian, *Various History*, T. Stanley (trans.), London, Thomas Basset, 1670.

Allbutt, Clifford, and Rolleston, Humphry Davy (eds.), *A System of Medicine*, Volume II, Part I, London, Macmillan and Co., Limited, 1908.

Ammianus Marcellinus, *The Chronicles of Events*, J. L. Rolfe (trans), The Loeb Classical Library, Cambridge, Harvard University Press, 1935.

Antiphanes, *Comicorum Atticorum Fragmenta*, Volumes I-II, R. Koch (ed.), Leipzig, B. G. Tuebner, 1884.

Apollodorus, *The Library*, H. G. Frazer (trans.), The Loeb Classical Library, London, Heinemann, 1921.

Apollonius Rhodius, *Argonautica*, R. C. Seaton (trans.), The Loeb Classical Library, New York, Macmillan, 1912.

Appian, *Roman History,* H. White (trans.), The Loeb Classical Library, London, Heinemann, 1912.

Apuleius, *Apologia*, H. E. Butler (trans.), Oxford, Clarendon Press, 1909.

Apuleius, *Metamorphoses*, W. Adlington (trans.), The Loeb Classical Library, London, Heinemann, 1915.

Arcamone, F., *et al.*, "Production of Lysergic Acid Derivatives by a Strain of *Claviceps paspali* Stevens and Hall in Submerged Culture," *Nature* [London], 187 (1960), pp. 238-239.

Archilochus, *Carmina; the Fragments of Archilochos*, G. Davenport (trans.), Berkeley, University of California Press, 1964.

Aristotle, *Metaphysics*, H. Tredennick (trans.), The Loeb Classical Library, London, Heinemann, 1933.

Aristotle, *On Marvellous Things Heard*, W. S. Hett (trans.), The Loeb Classical Library, Cambridge, Harvard University Press, 1934.

Arrian, *Anabasis of Alexander*, E. I. Robson (trans.), The Loeb Classical Library, Cambridge, Harvard University Press, 1929.

Athenaeus, *Deipnosophistae*, I. B. Gulick (trans.), The Loeb Classical Library, London, Heinemann, 1927.

Augustine, *De Civitate Dei*, Volumes I-II, J. Healey (trans.), New York, E. P. Dutton & Co., Inc., 1945.

Aurelianus, Caelius, *On Acute Diseases and On Chronic Diseases*, I. E. Drabkin (trans.), Chicago, University of Chicago Press, 1950.

Ausonius of Bordeaux, *Epigrams*, H. G. Evelyn-White (trans.) The Loeb Classical Library, London, Heinemann, 1919.

Ausonius of Bordeaux, *The Professors at Bordeaux*, H. G. Evelyn-White (trans.) The Loeb Classical Library, London, Heinemann, 1919.

Ausonius of Bordeaux, *Technopaegnion*, H. G. Evelyn-White (trans.) The Loeb Classical Library, London, Heinemann, 1919.

Bakels, C. C., "The contents of ceramic vessels in the Bactria-Margiana Archaeological Complex, Turkmenistan," Electronic Journal of Vedic Studies, 9 (2003).

Barger, George, *Ergot and Ergotism: A Monograph Based on the Dohme Lectures Delivered in Johns Hopkins University, Baltimore*, London, Gurney and Jackson, 1931.

Barger, George, "The Alkaloids of Ergot," *Handbuch der Experimentellen Pharmakologie*, Volume VI, W. Heubner and J. Schüller (eds.), Berlin, Julius Springer, 1938, pp. 84-245.

Bauschatz, Paul C., *The Well and the Tree: World and Time in Early Germanic Culture*, Amherst, University of Massachusetts Press, 1982.

Bhattacharyya, N. N., *History of the Tantric Religion: A Historical, Ritualistic and Philosophical Study*, New Delhi, Manohar, 1982.

Bloomfield, Maurice (trans.), *Hymns of the Atharva-Veda: Together with Extracts from the Ritual Books and the Commentaries*, Oxford, Clarendon Press, 1897.

Bodewitz, H. W. (trans.), *The Jyotiṣṭoma Ritual: Jaiminīya Brāhmaṃa I, 66-364*, Leiden, E. J. Brill, 1990.

Boëthius, *The Consolation of Philosophy*, H. F. Stewart (trans.), The Loeb Classical Library, London, Heinemann, 1918.

Bove, Frank James, *The Story of Ergot: For Physicians, Pharmacists, Nurses, Biochemists, Biologists and Others Interested in the Life Sciences*, Basel, S. Karger, 1970.

Brough, John, "Soma and *Amanita muscaria*," *Bulletin of the School of Oriental and African Studies*, 34 (1971), pp. 331-362.

Brough, John, "Problems of the «Soma-Mushroom» Theory," *Indologica Taurinensia*, 1 (1973), pp. 21-32.

Burrow, T., "The Early Āryans," *A Cultural History of India*, A. L. Basham (ed.), Oxford, Clarendon Press, 1975, pp. 20-29.

Caesar, Julius, *Gallic War*, H. J. Edwards (trans.), The Loeb classical Library, Cambridge, Harvard University Press, 1917.

Caland, W. (trans.), *Śāṅkhāyana-Śrautasūtra: being a major yājñika text of the Ṛgveda*, Nagpur, International Academy of Indian Culture, 1953.

Callimachus, *Aetia*, C. A. Trypanos (trans.), The Loeb classical Library, Cambridge, Harvard University Press, 1958.

Callimachus, *Hymn to Apollo*, G. H. Mair (trans.), *Hymns and Epigrams*, The Loeb classical Library, Cambridge, Harvard University Press, 1955.

Cameron, E. A., and French, E. B., "St. Anthony's Fire Rekindled: Gangrene due to Therapeutic Dose of Ergotamine," *British Medical Journal*, 2 (1960), p. 28.

Campbell, L. A., *Mithraic Iconography and Ideology*, Leiden, E. J. Brill, 1968.Caland, W. (trans.), *Pañcaviṃśa-Brāhmaṇa: The Brāhmaṇa of Twenty Five Chapters*, Calcutta, Asiatic Society of Bengal, 1931.

Chao, Jew-Ming, and Der Marderosian, Ara H., "Identification of Ergoline Alkaloids in the Genus *Argyreia* and Related Genera and their Chemotaxonomic Implications in the Convolvulaceae," *Phytochemistry*, 12 (1973), pp. 2435-2440.

Chao, Jew-Ming, and Der Marderosian, Ara H., "Ergoline Alkaloidal Constituents of Hawaiian Baby Wood Rose, *Argyreia nervosa* (Burm. f.) Bojer," *Journal of Pharmaceutical Sciences*, 62 (1973), pp. 588-591.

Cicero, *De Divinatione*, W. A. Falconer (trans.), The Loeb Classical Library, London, Heinemann, 1923.

Cicero, *De Legibus*, C. W. Keyes (trans.), The Loeb classical Library, Cambridge, Harvard University Press, 1959.

Claudian, *Against Eutropius*, M. Platnauer (trans.), The Loeb Classical Library, London, Heinemann, 1922.

Claudian, *Gothic War*, M. Platnauer (trans.), The Loeb Classical Library, London, Heinemann, 1922.

Clement of Alexandria, *Stromata*, J. Ferguson (trans.), Washington D.C., Catholic University of America Press, 1991.

Columella, Lucius Junius Moderatus, *On Agriculture and Trees*, E. S. Forster and E. H. Heffner (trans.), Volume III, Cambridge, Harvard University Press, 1955.

Cowell, E. B., and Thomas, F. W. (trans.), *The Harṣa-carita of Bāṇa*, Delhi, Motilal Banarsidass, 1961.

CRC Atlas of Spectral Data and Physical Constants for Organic Compounds, Volume III, Second Edition, J. G. Grasselli and W. M. Ritchey (eds.), Cleveland, CRC Press, Inc., 1975.

CRC Handbook of Data on Organic Compounds, Volume I, R. C. Weast and M. J. Astle (eds.), Boca Raton, Fl., CRC Press, Inc., 1985.

CRC Handbook of Microbiology, Volume V, Second Edition, A. I. Laskin and H. A. Lechevalier (eds.), Boca Raton, Fl., CRC Press, Inc., 1977.

Creighton, Charles, *A History of Epidemics in Britain*, Volume I, Cambridge, Cambridge University Press, 1891.

Dandekar, R. N. (trans.), *Śrautakośa: Encyclopædia of Vedic Sacrificial Ritual Comprising the Two Complementary Sections, Namely, the Sanskrit Section and the English Section*, Volume I, Part 2, Poona, Vaidika Saṁsodhana Maṇḍala, 1962.

Darby, William J., and Ghalioungui, Paul, and Grivetti, Louis, *Food: The Gift of Osiris*, Volume II, London, Academic Press, 1977.

Davidson, H. R. Ellis, "Mithras and Wodan," *Etudes Mithraiques*, 4 (1978), pp. 99-110.

Davidson, H. R. Ellis, *Myths and symbols in pagan Europe: Early Scan-*

dinavian and Celtic religions, Manchester, Manchester University Press, 1988.

Demosthenes, *Against Aristocrates*, J. H. Vince (trans.), The Loeb Classical Library, Cambridge, Harvard University Press, 1935.

Dictionary of Natural Products, Volume III, London, Chapman & Hall, 1994.

Dio Cassius, *Roman History*, Volumes I-IX, E. Cary (trans.), The Loeb Classical Library, London, Heinemann, 1914-1927.

Dio Chrysostom, *Discourses*, J. W. Cohoon (trans.), The Loeb Classical Library, London, Heinemann, 1932.

Diodorus Siculus, *The Library of History*, Volumes I-XII, C. H. Oldfather (trans.), The Loeb Classical Library, London, Heinemann, 1933-1967.

Diogenes Laertius, *Lives of Eminent Philosophers*, R. D. Hicks (trans.), The Loeb Classical Library, London, Heinemann, 1925.

Dionysius of Halicarnassus, *Roman Antiquities*, Volumes I-VII, E. Cary (trans.), The Loeb Classical Library, Cambridge, Harvard University Press, 1937-1950.

Dioscorides Pedanius of Anazarbos, *De Materia Medica*, J. Goodyear (trans.), Oxford, Oxford University Press, 1934.

Doniger O'Flaherty, Wendy, "The Post-Vedic History of the Soma Plant," in R. Gordon Wasson, *Soma: Divine Mushroom of Immortality*, New York, Harcourt Brace Jovanovich, Inc., 1968, pp. 95-147.

Doniger, Wendy, "'Somatic' Memories of R. Gordon Wasson," *The Sacred Mushroom Seeker: Essays for R. Gordon Wasson,* T. J. Riedlinger (ed.), Portland, Or., Dioscorides Press, 1990, pp. 55-59.

Dresden, M. J., "Mythology of Ancient Iran," *Mythologies of the Ancient World*, S. N. Kramer (ed.), New York, Doubleday, 1961, pp. 331-366.

Duchesne-Guillemin, Jacques, "The Wise Men from the East in the Western Tradition," *Papers in Honour of Professor Mary Boyce*, Volume I, Leiden, E. J. Brill, 1985, pp. 149-157.

Dumézil, Georges, "Le Festin d'Immortalité: Étude de Mythologie Comparée Indo-Européenne," *Annales du Musée Guimet*, 34 (1924), pp. 1-318.

Eggeling, Julius (trans.), *The Satapatha-Brâhmana: According to the Text of the Mâdhyandina School*, Volumes I-V, Oxford, Clarendon Press, 1882-1900.

Emboden, William, *Narcotic Plants*, New York, Macmillan Publishing Co.,

Inc., 1979.

Erdosy, George, "Ethnicity in the Rigveda and its Bearing on the Question of Indo-European Origins," *South Asian Studies*, 5 (1989), pp. 34-47.

Erdosy, George, "Language, material culture and ethnicity: Theoretical perspectives," *The Indo-Aryans of Ancient South Asia: Language, Material Culture and Ethnicity*, G. Erdosy (ed.), Berlin, Walter de Gruyter, 1995, pp. 1-31.

Euripides, *Alcestis*, A. S. Way (trans.), The Loeb Classical Library, Cambridge, Harvard University Press, 1912.

Euripides, *Helen*, A. S. Way (trans.), The Loeb Classical Library, Cambridge, Harvard University Press, 1912.

Euripides, *Madness of Heracles*, A. S. Way (trans.), The Loeb Classical Library, Cambridge, Harvard University Press, 1912.

Euripides, *Hippolytus*, A. S. Way (trans.), The Loeb Classical Library, Cambridge, Harvard University Press, 1912.

Euripides, *Ion*, A. S. Way (trans.), The Loeb Classical Library, Cambridge, Harvard University Press, 1912.

Euripides, *Medea*, A. S. Way (trans.), The Loeb Classical Library, Cambridge, Harvard University Press, 1912.

Euripides, *Rhesus*, A. S. Way (trans.), The Loeb Classical Library, Cambridge, Harvard University Press, 1912.

Eusebius, *Praeparatio Evangelica*, E. H. Gifford (trans.), Oxford, Claredon Press, 1975.

Eustathius, *Eustathii Archiepiscopi Thessalonicensis Commentarii ad Homeri Odysseum*, J. G. Stallbaum (trans.), Cambridge, Cambridge University Press, 2010.

Fajardo, J. E., *et al.*, "Retention of Ergot Alkaloids in Wheat During Processing," *Cereal Chemistry*, 72 (1995), pp. 291-298.

Falk, Harry, "Soma I and II," *Bulletin of the School of Oriental and African Studies*, 52 (1989), pp. 77-90.

Fanchamps, A., "Some Compounds with Hallucinogenic Activity," *Ergot Alkaloids and Related Compounds*, B. Berde and H. O. Schild (eds.), Berlin, Springer-Verlag, 1978, pp. 567-614.

Festi, Francesco, and Samorini, Giorgio, "*Claviceps paspali* and the Eleusinian *Kykeon*: A Correction," *Entheogen Review*, 8 (1999), pp. 96-97.

Firmicus Maternus, *Mathesis*, J. R. Bram (trans.), Park Ridge, Noyes Press, 1975.

Flam, Louis, "Fluvial Geomorphology of the Lower Indus Basin (Sindh, Pakistan) and the Indus Civilization," *Himalaya to the Sea: Geology, geomorphology and the Quaternary*, John F. Shroder, Jr. (ed.), London, Routledge, 1993, pp. 265-287.

Flattery, David Stophlet, and Schwartz, Martin, *Haoma and Harmaline: The Botanical Identity of the Indo-Iranian Sacred Hallucinogen "Soma" and its Legacy in Religion, Language, and Middle Eastern Folklore*, Berkeley, University of California Press, 1989.

Fol, A., and Marazov, I., *Thrace & the Thracians*, New York, St. Martin's Press, 1977.

Forbes, R. J., *Studies in Ancient Technology*, Volume III, Leiden, E. J. Brill, 1955.

Forni, Gaetano, "The Origin of Grape Wine: A Problem of Historical-Ecological Anthropology," *Gastronomy: The Anthropology of Food and Food Habits*, M. L. Arnott (ed.), The Hague, Mouton Publishers, 1975, pp. 67-78.

Gelder, Jeannette M. van (trans.), *The Mānava Śrautasūtra: belonging to the Maitrāyaṇī Saṃhitā*, New Delhi, International Academy of Indian Culture, 1963.

Gershevitch, Ilya, "An Iranianist's View of the Soma Controversy," *Mémorial Jean De Menasce*, A. Tafozzoli (ed.), Louvain, Imprimerie Orientaliste, 1974, pp. 45-75.

Glässer, A., "Some Pharmacological Actions of D-Lysergic Acid Methyl Carbinolamide," *Nature* [London], 189 (1961), pp. 313-314.

Gonda, Jan, *Rice and Barley Offerings in the Veda*, Leiden, E. J. Brill, 1987.

Griffith Ralph T. H. (trans.), *The Hymns of the Rgveda*, Second Edition, Volumes I-II, Benares, E. J. Lazarus, 1889-1892.

Griffith, Ralph T. H. (trans.), *The Texts of the White Yajurveda*, Benares, E. J. Lazarus and Co., 1899.

Griffith, Ralph T. H. (trans.), *The Hymns of the Atharva-Veda*, Second Edition, Volumes I-II, Benares, E. J. Lazarus & Co., 1916-1917.

Griffith, Ralph T. H. (trans.), *The Hymns of the Sāmaveda*, Fourth Edition, Varanasi, Chowkhamba Sanskrit Series Office, 1963.

Haug, Martin, *Essays on the Sacred Language, Writings, and Religion of the Parsis*, Popular Edition, London, Kegan Paul, Trench, Trübner & Co., 1883.

Heacock, R. A., "Psychotomimetics of the Convolvulaceae," *Progress in Medicinal Chemistry*, 11 (1975), pp. 91-118.

Heraclitus, *The Cosmic Fragments*, G. S. Kirk (trans.), Cambridge, Cambridge University Press, 1954.

Herodian, *History of the Roman Empire*, C. R. Whittaker (trans.), The Loeb Classical Library, Cambridge, Harvard University Press, 1969.

Herodotus, *Histories*, Volumes I-IV, A. D. Godley (trans.), The Loeb Classical Library, London, Heinemann, 1923.

Hesiod, *Theogony,* Volume I, G. W. Most (trans.), The Loeb Classical Library, Cambridge, Harvard University Press, 2006.

Hiebert, Fredrik T., *Origins of the Bronze Age Oasis Civilization in Central Asia*, Cambridge, Peabody Museum of Archaeology and Ethnology, Harvard University, 1994.

Hiebert, Fredrik T., "South Asia from a Central Asian perspective," *The Indo-Aryans of Ancient South Asia: Language, Material Culture and Ethnicity*, G. Erdosy (ed.), Berlin, Walter de Gruyter, 1995, pp. 192-205.

Hillebrandt, Alfred, *Vedic Mythology*, Volume I, S. R. Sarma (trans.), Delhi, Motilal Banarsidass, 1980.

Hippocrates, *Airs Waters Places*, W. H. S. Jones (trans.), The Loeb Classical Library, London, Heinemann, 1923.

Hippocrates, *Oath*, W. H. S. Jones (trans.), The Loeb Classical Library, London, Heinemann, 1923.

Hippolytus, *The Refutation of all Heresies*, A. Roberts and J. Donaldson (trans.), Grand Rapids, Eerdmans, 1951.

Hofmann, Albert, "The Active Principles of the Seeds of *Rivea corymbosa* and *Ipomoea violacea*," *Botanical Museum Leaflets, Harvard University*, 20 (1963), pp. 194-212.

Hofmann, Albert, *Die Mutterkornalkaloide*, Stuttgart, Ferdinand Enke, 1964.

Hofmann, Albert, "*Teonanácatl* and *Ololiuqui*, two ancient magic drugs of Mexico," *Bulletin on Narcotics*, 23 (1971), pp. 3-14.

Hogarth, David G., *Ionia and the East; Six Lectures Delivered before the University of London*, Oxford, Clarendon Press, 1909.

Holm, LeRoy G., *et al.*, *The World's Worst Weeds: Distribution and Biology*, Honolulu, University Press of Hawaii, 1977.

Homer, *Iliad*, Volumes I-II, A. T. Murray (trans.), The Loeb Classical Library, Cambridge, Harvard University Press, 1925.

Homer, *Odyssey*, Volumes I-II, A. T. Murray (trans.), The Loeb Classical Library, Cambridge, Harvard University Press, 1925.

Homeric Hymn to Demeter, H. P. Foley (trans.), Princeton, Princeton University Press, 1994.

Hopkins, E. Washburn, "Soma," *Encyclopædia of Religion and Ethics*, Volume XI, J. Hastings (ed.), Edinburgh, T. & T. Clark, 1920.

Horace, *Carmina*, W. G. Williams (trans.), Oxford, Claredon Press, 1969.

Horace, *Epistles*, H. R. Fairclough (trans.), The Loeb Classical Library, Cambridge, Harvard University Press, 1926.

Iamblichus, *Mysteries of the Egyptians, Chaldeans, and Assyrians*, T. Taylor (trans.), San Diego, Wizards Bookshelf, 1984.

Iamblichus, *De Vita Pythagorica*, C. Clark (trans.), Liverpool, Liverpool University Press, 1989.

Ingalls, Daniel H. H., "Soma" [Book Review], *The New York Times Book Review*, 5 September 1971, p. 15.

Ingalls, Daniel H. H., "Remarks on Mr. Wasson's *Soma*," *Journal of the American Oriental Society*, 91 (1971), pp. 169-191.

Isocrates, *Busiris*, L. V. Hook (trans.), The Loeb Classical Library, Cambridge, Harvard University Press, 1945.

Isocrates, *Panathenaicus* G. Norlin (trans.), The Loeb Classical Library, Cambridge, Harvard University Press, 1929.

Jarvik, M. E., *et al.*, "Comparative Subjective Effects of Seven Drugs including Lysergic Acid Diethylamide (LSD-25)," *Journal of Abnormal and Social Psychology*, 51 (1955), pp. 657-662.

Josephus, *Against Apion*, H. St. J. Thackeray (trans.), The Loeb Classical Library, Cambridge, Harvard University Press, 1926.

Joshi, K. L. (ed.), *Atharva-Veda Sa§hitā*, Volume I, Delhi, Parimal Publications, 2000.

Justin, *Epitome of the Philippic History of Pompeius Trogus*, J. C. Yardley (trans.), Atlanta, Scholars Press, 1994.

Kaegi, Adolf, *The Rigveda: The Oldest Literature of the Indians*, R. Arrowsmith (trans.), Boston, Ginn and Company, 1902.

Kane, Pandurang Vaman, *History of Dharmaśāstra (Ancient and Mediæval Religious and Civil Law)*, Volume II, Second Edition, Poona, Bhandarkar Oriental Research Institute, 1974.

Kapadia, B. H., *Soma in the Legends*, Vallabh Vidyanagar, B. H. Kapadia, 1958.

Kapadia, B. H., *A Critical Interpretation and Investigation of Epithets of Soma*, Vallabh Vidyanagar, B. H. Kapadia, 1959.

Kashikar, C. G., *Identification of Soma*, Pune, C. G. Kashikar, 1990.

Keith, Arthur Berriedale (trans.), *The Veda of the Black Yajus School entitled Taittiriya Sanhita*, Volumes I-II, Cambridge, Harvard University Press, 1914.

Keith, Arthur Berriedale (trans.), *Rigveda Brahmanas: The Aitareya and Kauṣītaki Brāhmaṇas of the Rigveda*, Cambridge, Harvard University Press, 1920.

Keith, Arthur Berriedale, *The Religion and Philosophy of the Veda and Upanishads*, Volume II, Cambridge, Harvard University Press, 1925.

Kenoyer, Jonathan Mark, "Interaction systems, specialised crafts and culture change: The Indus Valley Tradition and the Indo-Gangetic Tradition in South Asia," *The Indo-Aryans of Ancient South Asia: Language, Material Culture and Ethnicity*, G. Erdosy (ed.), Berlin, Walter de Gruyter, 1995, pp. 213-257.

Kerényi, C., *Eleusis: Archetypal Image of Mother and Daughter*, R. Manheim (trans.), New York, Bollingen Foundation, 1967.

Kerényi, C., *Dionysos: Archetypal Image of Indestructible Life*, R. Manheim (trans.), Princeton, Princeton University Press, 1976.

Kolhatkar, Madhavi Bhaskar, *Surā: The Liquor and the Vedic Sacrifice*, New Delhi, D. K. Printworld Ltd., 1999.

Krikorian, Abraham D., "Were the Opium Poppy and Opium Known in the Ancient Near East?," *Journal of the History of Biology*, 8 (1975), pp. 95-114

Kritikos, P. G., and Papadaki, S. P., "The history of the poppy and of opium and their expansion in antiquity in the eastern Mediterranean area," G. Michalopoulos (trans.), *Bulletin on Narcotics*, 19 (1967), pp. 5-10, 17-38.

La Barre, Weston, "The Deathless Gods," *The Ghost Dance: Origins of Reli-*

gion, New York, Delta Publishing Co. Inc., 1970.

Lal, B. B., "The Indus Civilization," *A Cultural History of India*, A. L. Basham (ed.), Oxford, Clarendon Press, 1975, pp. 11-19.

Leach, Edmund, "Aryan Invasions over Four Millennia," *Culture Through Time: Anthropological Approaches*, Emiko Ohnuki-Tierney (ed.), Stanford, Stanford University Press, 1990, pp. 227-245.

Liddell, Henry George, and Scott, Robert, *A Greek-English Lexicon*, Ninth Edition, Oxford, Clarendon Press, 1996.

Lucan, *Pharsalia*, J. D. Duff (trans.), The Loeb Classical Library, London, Heinemann, 1928.

Lucretius, *De Rerum Natura*, W. H. D. Rouse (trans.), The Loeb Classical Library, London, Heinemann, 1928.

Macdonell, Arthur Anthony, *Vedic Mythology*, Strassburg, Karl J. Trübner, 1897.

Macdonell, Arthur Anthony, and Keith, Arthur Berriedale, *Vedic Index of Names and Subjects*, Volumes I-II, London, John Murray & Co., 1912.

Macdonell, Arthur Anthony, *A Vedic Reader for Students*, Oxford, Clarendon Press, 1917.

Macdonell, Arthur Anthony (trans.), *Hymns from the Rigveda*, London, Oxford University Press, 1922.

Macrobius, *Saturnalia*, P. V. Davies (trans.), New York, Columbia University Press, 1969.

Mallory, J. P., and Adams, D. Q., *The Oxford Introduction to Proto-Indo-European and the Proto-Indo-European World*, Oxford, Oxford University Press, 2006.

Marshall, Sir John (ed.), *Mohenjo-daro and the Indus Civilization...*, Volume I, London, Arthur Probsthain, 1931.

Megesthenes, *Indika*, E. A. Schwanbeck (trans.), Calcutta, Thackery, Spink, 1877.

Merlin, Mark David, *On the Trail of the Ancient Opium Poppy*, Rutherford, Fairleigh Dickinson University Press, 1984.

Merrillees, Robert S., "Opium Trade in the Bronze Age Levant," *Antiquity*, 36 (1962), pp. 287-292.

Merrillees, Robert S., "Opium again in Antiquity," *Levant*, 11 (1979), pp. 167-171.

Merrillees, Robert S., "Highs and Lows in the Holy Land: Opium in Biblical Times," *Eretz-Israel*, 20 (1988-1989), pp. 148-155.

Meyer-Melikyan, N. R., and Avetov, N. A., "Analysis of Floral Remains in the Ceramic Vessel from the Gonur Temenos," in Victor I. Sarianidi, *Margiana and Protozoroastrism*, Inna Sarianidi (trans.), Athens, Kapon Editions, 1998, Appendix I, pp. 176-177.

Meyer-Melikyan, N. R., "Analysis of Floral Remains from Togolok-21," in Victor I. Sarianidi, *Margiana and Protozoroastrism*, Inna Sarianidi (trans.), Athens, Kapon Editions, 1998, Appendix II, pp. 178-179.

Misra, V. N., "Climate, a Factor in the Rise and Fall of the Indus Civilization – Evidence from Rajasthan and Beyond," *Frontiers of the Indus Civilization*, B. B. Lal and S. P. Gupta (eds.), New Delhi, Indian Archaeological Society, pp. 461-489.

Mitra, Rájendralála, "Spirituous Drinks in Ancient India," *Journal of the Royal Asiatic Society of Bengal*, 42 (1873), pp. 1-23.

Müller, F. Max, *Biographies of Words and The Home of the Aryas*, London, Longmans, Green, and Co., 1888.

Müller, F. Max (trans.), *Vedic Hymns*, Volume I, Oxford, Clarendon Press, 1891.

Nesbitt, Mark, "Plant Use in the Merv Oasis," *Iran: Journal of the British Institute of Persian Studies*, 35 (1997), pp. 29-31.

Neuburger, Albert, *The Technical Arts and Sciences of the Ancients*, H. L. Brose (trans.), New York, Macmillan Company, 1930.

Nicander of Colophon, *Poems and Poetical Fragments*, A. S. F. Gow and A. F. Schofield (trans. and eds.), Cambridge, Cambridge University Press, 1953.

Nonnos, *Dionysiaca*, W. H. D. Rouse (trans.), The Loeb Classical Library, Cambridge, Harvard University Press, 1940.

Nooten, Barend A. Van, and Holland, Gary B. (eds.), *Rig Veda: A Metrically Restored Text with an Introduction and Notes*, Cambridge, Department of Sanskrit and Indian Studies, Harvard University, 1994.

Norman, Kenneth R., Review of "The Coming of the Aryans to Iran and India and the Cultural and Ethnic Identity of the Dāsas," *Acta Orientalia*, 51 (1990), pp. 288-296.

Norman, Kenneth R., "Dialect variation in Old and Middle Indo-Aryan," *The*

Indo-Aryans of Ancient South Asia: Language, Material Culture and Ethnicity, G. Erdosy (ed.), Berlin, Walter de Gruyter, 1995, pp. 278-292.

Nyberg, Harri, "The problem of the Aryans and the Soma: The botanical evidence," *The Indo-Aryans of Ancient South Asia: Language, Material Culture and Ethnicity*, G. Erdosy (ed.), Berlin, Walter de Gruyter, 1995, pp. 382-406.

Oldenberg, Hermann (trans.), *Vedic Hymns*, Volumes I-II, Oxford, Clarendon Press, 1897.

Oppian, *Cynegetica*, A. W. Mair (trans.), The Loeb Classical Library, London, Heinemann, 1928.

Orphic Hymns, The, A. N. Athanassakis (trans.), Missoula, Scholars Press for the Society of Biblical Literature, 1977.

Ott, H., and Hofmann, A., and Frey, A. J., "Acid-Catalyzed Isomerization in the Peptide Part of Ergot Alkaloids," *Journal of the American Chemical Society*, 88 (1966), pp. 1251-1256.

Ovid, *Fasti*, J. G. Frazer (trans.), The Loeb Classical Library, London, Heinemann, 1931.

Ovid, *Metamorphoses*, F. J. Miller (trans.), The Loeb Classical Library, London, Heinemann, 1916.

Pandurangi, K. T. (trans.), *Chandogyopanishad*, Chirtanur, Sriman Madhua Siddhantonnahini Sabha, 1987.

Paris Magic Papyrus, Bibliotheque nationale, France, Supplement grec 574, leaves 1-36.

Parpola, Asko, "The Coming of the Aryans to Iran and India and the Cultural and Ethnic Identity of the Dāsas," *Studia Orientalia*, 64 (1988), pp. 195-302.

Parpola, Asko, "Bronze Age Bactria and Indian Religion," *Studia Orientalia*, 70 (1993), pp. 81-87.

Parpola, Asko, "The problem of the Aryans and the Soma: Textual-linguistic and archaeological evidence," *The Indo-Aryans of Ancient South Asia: Language, Material Culture and Ethnicity*, G. Erdosy (ed.), Berlin, Walter de Gruyter, 1995, pp. 353-381.

Pausanias, *Description of Greece*, Volumes I-V, W. H. S. Jones and H. A. Ormerod (trans.), The Loeb Classical Library, London, Heinemann, 1925.

Penny Cyclopœdia of the Society for the Diffusion of Useful Knowledge, The,

Volume XIV, London, Charles Knight and Co., 1839.

Phanocles, *Commentary on Phanocles*, K. Alexander (trans.), Amsterdam, A. M. Hakkert, 1988.

Philo, *Every Good Man is Free*, F. H. Colson (trans.) The Loeb Classical Library, Cambridge, Harvard University Press, 1941.

Philostratus, Flavius, *On Heroes*, J. K. B. Maclean and E. B. Aitken (trans.), Atlanta, Society of Biblical Literature, 2002.

Piggott, Stuart, "A Late Bronze Age Wine Trade?" *Antiquity*, 33 (1959), pp. 122-123.

Pindar, *Nemean Odes*, J. Sandys (trans.), The Loeb Classical Library, Cambridge, Harvard University Press, 1915.

Pindar, *Pythian Odes*, J. Sandys (trans.), The Loeb Classical Library, Cambridge, Harvard University Press, 1915.

Plato, *Cratylus*, H. N. Fowler (trans.), The Loeb Classical Library, Cambridge, Harvard University Press, 1926.

Plato, *Laws*, R. G. Bury (trans.), The Loeb Classical Library, Cambridge, Harvard University Press, 1926.

Plato, *Phaedrus*, H. N. Fowler (trans.), The Loeb Classical Library, Cambridge, Harvard University Press, 1914.

Plato, *Protagoras*, W. R. M. Lamb (trans.), The Loeb Classical Library, Cambridge, Harvard University Press, 1924.

Plautus, Titus Maccius, *Amphitryon*, P. Nixon (trans.), The Loeb Classical Library, Cambridge, Harvard University Press, 1916.

Pliny, *Natural History*, H. Rackman (trans.), The Loeb Classical Library, Cambridge, Harvard University Press, 1938.

Plutarch, *Alexander*, B. Perrin (trans.), The Loeb Classical Library, Cambridge, Harvard University Press, 1919.

Plutarch, *Caesar*, B. Perrin (trans.), The Loeb Classical Library, Cambridge, Harvard University Press, 1919.

Plutarch, *Coriolanus*, B. Perrin (trans.), The Loeb Classical Library, Cambridge, Harvard University Press, 1916.

Plutarch, *De Defectu Oraculorum*, F. C. Babbitt (trans.) The Loeb Classical Library, Cambridge, Harvard University Press, 1936.

Plutarch, *Isis and Osiris*, F. C. Babbitt (trans.) The Loeb Classical Library, Cambridge, Harvard University Press, 1936.

Plutarch, *Lucullus*, B. Perrin (trans.), The Loeb Classical Library, London, Heineman, 1918.

Plutarch, *Moralia*, Volumes I-XV, F. C. Babbitt (trans.), The Loeb Classical Library, Cambridge, Harvard University Press, 1955-1957.

Plutarch, *Theseus*, B. Perrin (trans.), The Loeb Classical Library, Cambridge, Harvard University Press, 1914.

Porphyry, *Abstinance from Animal Food*, T. Taylor (trans.), London, Centaur Press, 1965.

Porphyry, *On the Cave of the Nymphs*, R. Lamberton (trans.), Barrytown, N.Y., Station Hill Press, 1983.

Prakash, Om, *Food and Drinks in Ancient India (From Earliest Times to c. 1200 A.D.)*, Delhi, Munshi Ram Manohar Lal, 1961.

Proclus, *Commentary on the Pythagorean Golden Verses*, N. Linley (trans.), Dissertation, State University of New York at Buffalo, 1984.

Procopius, *History of the Wars*, H. B. Dewing (trans.), The Loeb Classical Library, Cambridge, Harvard University Press, 1914.

Procopius, *Secret History*, H. B. Dewing (trans.), The Loeb Classical Library, Cambridge, Harvard University Press, 1914.

Procopius, *Buildings*, H. B. Dewing (trans.), The Loeb Classical Library, Cambridge, Harvard University Press, 1914.

Propertius, Sextus, H. E. Butler (trans.), The Loeb Classical Library, London, Heinemann, 1912.

Pythagorean Golden Verses, The, Johan C. Thom (trans.), Leiden, E. J. Brill, 1995.

Quintas Curtius, *History of Alexander*, J. C. Rolfe (trans.), The Loeb Classical Library, Cambridge, Harvard University Press, 1946.

Quintilian, *De Institutione Oratoria*, H. E. Butler (trans.), The Loeb Classical Library, London, Heinemann, 1921.

Quintus Smyrnaeus, *The Fall of Troy*, A. S. Way (trans.), The Loeb Classical Library, London, Heinemann, 1913.

Ranade, H. G. (trans.), *Kātyāyana Śrauta Sūtra: Rules for the Vedic Sacrifices*, Pune, H. G. Ranade and R. H. Ranade, 1978.

Ranade, H. G. (trans.), *Āśvalāyana Śrauta-Sūtram*, Volume I, Poona, R. H. Ranade, 1981.

Regino of Prüm, *De Harmonica Institutione*, M. P. LeRoux (trans.), Dissertation, Catholic University of America, 1965.

Řeháček, Zdeněk, and Sajdl, Přemysl, *Ergot Alkaloids: Chemistry, Biological Effects, Biotechnology*, Amsterdam, Elsevier, 1990.

Renfrew, J. M., "The archaeological evidence for the domestication of plants: methods and problems," *The domestication and exploitation of plants and animals*, P. J. Ucko and G. W. Dimbleby (eds.), Chicago, Aldine Publishing Company, 1969, pp. 149-172.

Riedlinger, Thomas J., "Wasson's Alternative Candidates for Soma," *Journal of Psychoactive Drugs*, 25 (1993), pp. 149-156.

Ruck, Carl A. P., and Bigwood , Jeremy, and Danny, Staples, and Ott, Jonathan, and Wasson, R. Gordon, "Entheogens," *Journal of Psychedelic Drugs*, 11 (1979), pp. 145-146.

Ruck, Carl A. P., "The Wild and the Cultivated: Wine in Euripides' *Bacchae*," *Journal of Ethnopharmacology*, 5 (1982), pp. 231-270.

Rhys, John, *Lectures on the Origin and Growth of Religion as Illustrated by Celtic Heathendom*, Oxford, Williams and Norgate, 1898.

Sappho

Sarianidi, Victor I., "South-west Asia: Migrations, the Aryans and Zoroastrians," *International Association for the Study of the Cultures of Central Asia, Information Bulletin*, 13 (1987), pp. 44-56.

Sarianidi, Victor I., "The Bactrian Pantheon," *International Association for the Study of the Cultures of Central Asia, Information Bulletin*, 10 (1987), pp. 5-20.

Sarianidi, Victor I., "Margiana and the Indo-Iranian world," *South Asian Archaeology 1993: Proceedings of the Twelfth International Conference of the European Association of South Asian Archaeologists held in Helsinki University 5-9 July 1993*, Volumes I-II, A. Parpola and P. Koskikallio (eds.), Helsinki, Soumalainen Tiedeakatemia, 1994, vol. II, pp. 667-680.

Sarianidi, Victor I., "Temples of Bronze Age Margiana: traditions of ritual architecture," F. Hiebert (trans.), *Antiquity*, 68 (1994), pp. 388-397.

Sarianidi, Victor I., "New Discoveries at Ancient Gonur," *Ancient Civilizations from Scythia to Siberia*, 2 (1995), pp. 289-310.

Sarianidi, Victor I., *Margiana and Protozoroastrism*, Inna Sarianidi (trans.), Athens, Kapon Editions, 1998.

Sarianidi, Victor I., "Margiana and Soma-Haoma," Electronic Journal of Vedic Studies, 9 (2003).

Schultes, Richard Evans, and Hofmann, Albert, *The Botany and Chemistry of Hallucinogens*, Springfield, Ill., Charles C. Thomas, 1973.

Scott, P. M., *et al.*, "Ergot Alkaloids in Grain Foods Sold in Canada," *Journal of AOAC International*, 75 (1992), pp. 773-779.

Seneca, *Epistulae Morales*, R. M. Gummere (trans.), The Loeb Classical Library, Cambridge, Harvard University Press, 1920.

Sennert, Daniel, *Of Agues and Fevers. Their Differences, Signes, and Cures. Divided into four Books: Made English by* N. D. B. M. *late of* Trinity *Colledge in* Cambridge, London, Printed by *J. M.* for *Lodowick Lloyd*, at the Castle in Cornhil, 1658.

Shaffer, Jim G., and Lichtenstein, Diane A., "The concepts of 'cultural tradition' and 'palaeoethnicity' in South Asian archaeology," *The Indo-Aryans of Ancient South Asia: Language, Material Culture and Ethnicity*, G. Erdosy (ed.), Berlin, Walter de Gruyter, 1995, pp. 126-154.

Shelley, William Scott, *The Elixir: An Alchemical Study of the Ergot Mushrooms*, Notre Dame, Ind., Cross Cultural Publications, Inc., 1995.

Shelley, William Scott, *The Origins of the Europeans: Classical Observations in Culture and Personality*, San Francisco, International Scholars Publications, 1998.

Sherratt, Andrew G., "Cups That Cheered," *Bell Beakers of the Western Mediterranean: Definition, Interpretation, Theory and New Site Data*, William H. Waldren and Rex Claire Kennard (eds.), Oxford, B.A.R., 1987, pp. 81-114.

Sherratt, Andrew G., "Sacred and Profane Substances: the Ritual Use of Narcotics in Later Neolithic Europe," *Sacred and Profane: Proceedings of a Conference on Archaeology, Ritual and Religion. Oxford, 1989*, P. Garwood, D. Jennings, R. Skeates and J. Toms (eds.), Oxford, Published by the Oxford Committee for Archaeology, 1991, pp. 50-64.

Silius Italicus, *Punica*, Volume I, J. D. Duff (trans.), The Loeb Classical Library, Harvard, Harvard University Press, 1983.

Skjærvø, P. Oktor, "The Avesta as source for the early history of the Iranians," *The Indo-Aryans of Ancient South Asia: Language, Material Culture and Ethnicity*, G. Erdosy (ed.), Berlin, Walter de Gruyter, 1995, pp. 155-176.

Slater, Gilbert, *The Dravidian Element in Indian Culture*, London, Ernest

Benn Limited, 1924.

Smith, Sydney, and Timmis, Geoffrey Millward, "The Alkaloids of Ergot. Part III. Ergine, a New Base obtained by the Degradation of Ergotoxine and Ergotinine," *Journal of the Chemical Society* [London], 1932, pp. 763-766.

Smith, Sydney, and Timmis, Geoffrey Millward, "The Alkaloids of Ergot. Part VI. Ergometrinine," *Journal of the Chemical Society* [London], 1936, pp. 1166-1169.

Smith, Sydney, and Timmis, Geoffrey Millward, "The Alkaloids of Ergot. Part VII. isoErgine and isoLysergic Acids," *Journal of the Chemical Society* [London], 1936, pp. 1440-1444.

Smith, Sydney, and Timmis, Geoffrey Millward, "The Alkaloids of Ergot. Part VIII. New Alkaloids of Ergot: Ergosine and Ergosinine," *Journal of the Chemical Society* [London], 1937, pp. 396-401.

Statius, *Thebaid*, J. H. Mozley (trans.), The Loeb Classical Library, Cambridge, Harvard University Press, 1955.

Steblin-Kamenskij, I., Review of *Papers in honour of Professor Mary Boyce*, *Bulletin of the School of Oriental and African Studies*, 50 (1987), pp. 376-378.

Stein, Aurel, "On the Ephedra, the Hūm Plant, and the Soma," *Bulletin of the School of Oriental and African Studies*, 6 (1931), pp. 501-514.

Stolaroff, Myron J., *The Secret Chief*, Charlotte, N.C., Multidisciplinary Association for Psychedelic Studies, 1997.

Stoll, A., and Hofmann, A., "The Ergot Alkaloids," *The Alkaloids: Chemistry and Physiology*, Volume VIII, R. H. F. Manske (ed.), New York, Academic Press, 1965, pp. 725-779.

Strabo, *Geographica*, Volumes I-VIII, H. L. Jones (trans.), The Loeb Classical Library, Cambridge, Harvard University Press, 1917-1932.

Synesius, *Oration*,

Tacitus, Cornelius, *The Annals*, J. Jackson (trans.), The Loeb Classical Library, Cambridge, Harvard University Press, 1931.

Tacitus, Cornelius, *Histories*, C. H. Moore (trans.), The Loeb Classical Library, Cambridge, Harvard University Press, 1931.

Tétényi, Péter, "Opium Poppy (*Papaver somniferum*): Botany and Horticulture," *Horticultural Reviews*, 19 (1997), pp. 373-408.

Theophrastus, *History of Plants*, Volumes I-II, Sir A. Holt (trans.), The Loeb Classical Library, London, Heinemann, 1919.

Thompson, R. Campbell, *A Dictionary of Assyrian Botany*, London, British Academy, 1949.

Thucydides, *History of the Peloponnesian War*, Volumes I-IV, C. F. Smith (trans.), The Loeb Classical Library, London, Heinemann, 1919.

Tryphiodorus, *Ilios*, A. W. Mair (trans.), The Loeb Classical Library, Cambridge, Harvard University Press, 1928.

Valerius Flaccus, *Argonautica*, J. H. Mozley (trans.), The Loeb Classical Library, Cambridge, Harvard University Press, 1934.

Varro, *De Lingua Latina*, Volumes I-II, R. G. Kent (trans.), The Loeb Classical Library, Cambridge, Harvard University Press, 1938.

Vats, Madho Sarup, *Excavations at Harappā...*, Volume I, Varanasi, Bhartiya Publishing House, 1974.

Velleius Paterculus, *Roman History*, F. W. Shipley (trans.), The Loeb Classical Library, Cambridge, Harvard University Press, 1924.

Virgil, *Aenid*, H. R. Fairclough (trans.), The Loeb Classical Library, Cambridge, Harvard University Press, 1916.

Virgil, *Ecologues*, H. R. Fairclough (trans.), The Loeb Classical Library, Cambridge, Harvard University Press, 1916.

Vitruvius Pollio, *On Architecture*, Volumes I-II, F. Granger (trans.), The Loeb Classical Library, London, Heinemann, 1914.

Wasson, R. Gordon, "Soma: the Divine Mushroom of Immortality," *Discovery*, 3 (1967), pp. 41-48.

Wasson, R. Gordon, *Soma: Divine Mushroom of Immortality*, New York, Harcourt Brace Jovanovich, Inc., 1968.

Wasson, R. Gordon, "*Soma*: Comments Inspired by Professor Kuiper's Review," *Indo-Iranian Journal*, 12 (1970), pp. 286-298.

Wasson, R. Gordon, "Soma of the Aryans: an ancient hallucinogen?" *Bulletin on Narcotics*, 22 (1970), pp. 25-30.

Wasson, R. Gordon, "The Soma of the Rig Veda: What Was It?" *Journal of the American Oriental Society*, 91 (1971), p. 169.

Wasson, R. Gordon, *Soma and the Fly-Agaric: Mr. Wasson's Rejoinder to Professor Brough*, Cambridge, Botanical Museum of Harvard University, 1972.

Wasson, R. Gordon, and Hofmann, Albert, and Ruck, Carl A. P., *The Road to Eleusis: Unveiling the Secret of the Mysteries*, New York, Harcourt Brace Jovanovich, Inc., 1978.

Wasson, R. Gordon, *The Wondrous Mushroom: Mycolatry in Mesoamerica*, New York, McGraw-Hill, 1980.

Wasson, R. Gordon, "The Last Meal of Buddha," *Journal of the American Oriental Society*, 102 (1982), pp. 591-603.

Wasson, R. Gordon, and Kramrisch, Stella, and Ott, Jonathan, and Ruck, Carl A. P., *Persephone's Quest: Entheogens and the Origins of Religion*, New Haven, Yale University Press, 1986.

Watkins, Calvert, "Let Us Now Praise Famous Grains," *Proceedings of the American Philosophical Society*, 122 (1978), pp. 9-17.

Watkins, Calvert (ed.), *The American Heritage Dictionary of Indo-European Roots*, Second Edition, Boston, Houghton Mifflin Company, 2000.

Webster, Peter, and Perrine, Daniel M., and Ruck, Carl A. P., "Mixing the *Kykeon*," *Eleusis*, New Series, 4 (2000), pp. 55-86.

Wheeler, Sir Mortimer, *The Indus Civilization; Supplementary Volume I*, Cambridge, Cambridge University Press, 1953.

Whitney, William Dwight (trans.), *Atharva-Veda Saṁhitā*, Volumes I-II, Cambridge, Harvard University, 1905.

Witzel, Michael, "Early Indian history: Linguistic and textual parametres," *The Indo-Aryans of Ancient South Asia: Language, Material Culture and Ethnicity*, G. Erdosy (ed.), Berlin, Walter de Gruyter, 1995, pp. 85-125.

Witzel, Michael, "Ṛgvedic history: poets, chieftains and polities," *The Indo-Aryans of Ancient South Asia: Language, Material Culture and Ethnicity*, G. Erdosy (ed.), Berlin, Walter de Gruyter, 1995, pp. 307-352.

Xenophon, *Anabasis*, L. L. Brownson (trans.), The Loeb Classical Library, Cambridge, Harvard University Press, 1922.

Xenophon, *Cyropaedia*, Volumes I-II, W. Miller (trans.), The Loeb Classical Library, London, Heinemann, 1914.

Xenophon, *Hellenica*, Volume I, C. L. Brown (trans.), The Loeb Classical Library, London, Heinemann, 1922.

Xenophon, *Memorabilia*, E. C. Marchant (trans.), The Loeb Classical Library, Cambridge, Harvard University Press, 1923.

Xenophon, *Symposium*, O. J. Todd (trans.), The Loeb Classical Library, Cambridge, Harvard University Press, 1923.

Zaehner, R. C., "The fortifying fungus" [Book Review], *The Times Literary Supplement*, 22 May 1969, p. 562.

Zohary, Daniel, and Hopf, Maria, *Domestication of Plants in the Old World: The origin and spread of cultivated plants in West Asia, Europe and the Nile Valley*, Third Edition, Oxford, Oxford University Press, 2000.

Printed in the United States
By Bookmasters